W9-BVT-506

BEYOND BANNEKER

BEYOND BANNEKER

BLACK MATHEMATICIANS AND THE PATHS TO EXCELLENCE

Erica N. Walker

Published by State University of New York Press, Albany

Printed in the United States of America

For information, contact State University of New York Press, Albany, NY
www.sunypress.edu

Production by Jenn Bennett
Marketing by Michael Campochiaro

Library of Congress Cataloging-in-Publication Data

Walker, Erica N.
Beyond Banneker : Black Mathematicians and the Paths to Excellence / Erica N. Walker.
pages cm
Includes bibliographical references and index.
ISBN 978-1-4384-5215-9 (hardcover : alk. paper) 1. African American mathematicians.
2. African American mathematicians—Education (Graduate) 3. Doctor of philosophy degree—United States. I. Title.
QA28.W35 2014
510.89'96073--dc23

2013029950

10 9 8 7 6 5 4 3 2 1

This book is dedicated to my parents—in memory
of my father, John T. Walker, Jr., and with gratitude
to my mother, Iona S. Walker. Their influence, guidance,
and love of learning have made me the person I am.

CONTENTS

PREFACE

"The Substance of Things Hoped For, the Evidence of Things Not Seen"

> Thomas Fuller, known as the Virginia Calculator, was stolen from his native Africa at the age of fourteen. . . . When he was about seventy years old, two gentlemen . . , having heard . . . of his extraordinary powers of arithmetic, sent for him and had their curiosity sufficiently gratified by the answers which he gave. . . . In 1789 he died at the age of 80 years, having never learned to read or write, in spite of his extraordinary power of calculation.
>
> —E. W. Scripture, *Arithmetical Prodigies* (1891, p. 3)

Thomas Fuller's (1709–1789) largely unknown life stands as an unfortunate record of limits placed on a potential mathematical genius. But for his birth into a free American family, the mathematical contributions of the better-known Benjamin Banneker (1731–1806), also of African descent, would similarly be lost to history. Fuller's anonymity and Banneker's relative fame stand as a commentary on the obstacles and opportunities that have circumscribed Black American mathematical talent across three centuries.

These two men—Fuller and Banneker—are the first recognized mathematical personages in the United States of African descent. It is practically impossible to name another U.S.-born Black mathematician until Elbert Frank Cox, who was the first Black person to earn his doctorate in mathematics in the United States, in 1925 (from Cornell University). It took nearly 20 years after Cox's achievement before the first African American woman, Euphemia Lofton Haynes, earned her doctorate in mathematics, in 1943, from Catholic University. The obstacles that Banneker, Fuller, and others faced in demonstrating their mathematical potential and talent centered primarily around the second-class status ascribed to them in the United States due to their racial heritage. But there was opportunity as well, even within the rigid confines of the segregated era. At times, opportunity was crafted and cultivated so that they could surmount obstacles that had nothing to do with merit but everything to do with race. Sometimes opportunity, fleeting and astonishingly present, was seized—a matter of being in the right place at the right time.

What may be surprising about the Black mathematicians' narratives shared in this book is that many of the elements in Fuller's and Banneker's 18th-century mathematical lives—talent unrecognized and sometimes unrewarded, as well as the sometimes startling and serendipitous nature of opportunity—resonate throughout the lives of contemporary Black mathematicians, young and old. History—the lived experiences of Black Americans in the United States as slaves, as free persons, as second-class and finally equal citizens under the law—plays an indelible role in shaping the experiences of Black mathematicians. This shared history and how it manifests itself in the formative, educational, and professional experiences of Black mathematicians is reflected in *how* they come to do and practice mathematics (Walker, 2009, 2011). This is not to say that the mathematics in which Black mathematicians engage is a special kind of mathematics unique to their ethnic heritage but rather that for many, how they conceptualize their professional identities and communities is in large part based on their experiences of being both Black and mathematically talented.

When told of this project exploring the experiences of Black mathematicians, defined for this book as those who have earned their PhDs in a mathematical science, a colleague in educational research asked "Are there any?" That this question can be asked with some sincerity begins to suggest the very real racialized space(s) that Black mathematicians occupy in a supposedly color-blind discipline and perhaps explains why Benjamin Banneker was compelled more than two centuries ago in 1791 to write Thomas Jefferson a letter exhorting him, at length, to do what he could to ameliorate the stereotypes about African Americans' intellectual capacity[1]:

> I apprehend you will readily embrace every opportunity to eradicate that train of absurd and false ideas and opinions which so generally prevail with respect to us. (Bedini, 1999, p. 158)

Banneker's appeal to Jefferson to use his authority and power to combat the (mis)representations of African Americans' intellectual ability finds a counterpoint two centuries later in the writings of the 10th Black American to earn his PhD in mathematics, Wade Ellis, Sr.:

> As a people, we have had more than our share of the academically hereditary disaffection all peoples seem to experience relative to mathematics. . . . Nowadays our promising youth are even more menacingly threatened by exposure to teachers . . . who have been convinced to their very viscera that Blacks . . . are abysmally and irrevocably hopeless as far as mathematics is concerned. (Newell, Gipson, Rich, & Stubblefield, 1980, p. ix)

Ellis's words appear in the first volume of its kind devoted to Black mathematicians and their research.[2] Like Banneker, Ellis (1909–1989) (who earned his PhD in mathematics from the University of Michigan in 1944 and was the first Black mathematician to work as a faculty member in a predominantly White institution, Oberlin College) was concerned that Blacks—specifically, Black youths—might be hampered by others' limited perceptions of their intellectual potential and ability. Both Ellis and Banneker would presume that mathematical talent among African Americans is not singular and rare.

This book answers questions about uniqueness and existence. But beyond providing answers for these questions, it demonstrates the power of history, place, and community in cultivating mathematics excellence and identity. It explores how Black mathematicians combat negative perceptions of their mathematics ability. It also examines how they form communities that support and contribute to mathematical knowledge and how these communities occupy multiple spaces within formal and informal educational contexts. It links the practices of some mathematicians to those of their forebears and honors the work and lives of those who, like Thomas Fuller, are woefully undersung.

ACKNOWLEDGMENTS

I wish to acknowledge, with immense gratitude, all of the mathematicians who participated in this study and generously shared their important stories with me. Their memories, histories, and experiences were so rich that a journal article was not enough to capture their mathematical lives, and although I fear this book is also too limited a forum for their narratives, I hope they believe it does them justice. I especially thank Drs. Clarence Stephens, Evelyn Granville, Scott Williams, Sylvia Bozeman, Bill Massey, Mel Currie, Johnny Houston, and Duane Cooper, as well as Drs. Ron Mickens and Fred Bowers, for their active encouragement, their advice, and their support of this project. I am particularly appreciative of Dr. David Blackwell, who spoke with me early on but sadly passed away before I finished this book. All were tremendously helpful; any errors in this work are my own.

This book could not have been completed without the support of Teachers College colleagues. In particular, I thank Professor Fran Schoonmaker for her interest and support of my work with high school mathematics students and in addition, for suggesting several years ago when I told her about the very interesting interviews with mathematicians I was conducting, that I had a book here. I also thank the provost and dean of Teachers College, Thomas James, and the Mathematics, Science, and Technology Department chairperson, O. Roger Anderson, for their critical support of this project.

I am so grateful to all of my colleagues in the Program in Mathematics: they have always supported and valued my work in multiple ways. This is a rare gift, and I appreciate their interest in this project. Thanks especially to Drs. Henry Pollak, Bruce Vogeli, Alexander Karp, and H. Philip Smith for many consultations, conversations, and connections to mathematicians, as well as TC faculty colleagues Anna Neumann and Aaron Pallas for their support of this work. Henry Pollak especially was invaluable, as he facilitated two important connections that led to the genesis of this book.

Our students in the Program in Mathematics are top-notch, and I gratefully acknowledge the key research assistance of Dr. Viveka Borum, Nathan Alexander, Dr. Julie Solomon Patterson, A. J. Stachelek, Dr. Heather Gould, Nicole Fletcher, and Simone Salmon. Although all of them did not work on this project, their support of my work on this and other projects freed my time and allowed me to work on this book uninterrupted and without distraction at critical moments. A very special thanks to our program secretary, Ms. Krystle Hecker, whose support was a great help and who has the additional gift of being a consummate mind reader.

Very special thanks to my wonderful friends Adey Stembridge; Alicia Dickerson; Alissa Gardenhire; Anu Cairo; Brett Felder; Darlene Martin; Denise Ross; Dorothy White; Elvira Prieto; Hal Smith; Helene Liss; Joe Rogers, Jr.; Jolene Lane Lewis; Kobi Abayomi; Maisha Fisher Winn; Maria Ledesma; Melissa Chabran; Robyn, Bill, and William Brady Ince; Saeyun Lee; Samona Tait; Tanya Odom; Toni Miranda; and Tracy Thomas for their interest in this work, for helpful suggestions, for travel/writing breaks, for pep talks and writing café visits, and for being willing to talk through ideas related to this book ad nauseum for the last several years.

I can't thank my TC colleagues and good friends Arshad Ali, Cally Waite, Lalitha Vasudevan, Lesley Bartlett, and Sandra Okita enough for all of the writing sessions and retreats, especially over the last year when I needed a final push. There's something to be said for gentle but persistent peer pressure. I am also grateful to my teachers, professors, and mentors for their support and inspiration, especially Charles Willie, Dick Murnane, Leah McCoy, Valerie Kimbrough, Yvonne Pringle, and Jacqueline Leonard.

Finally, I couldn't have done this work without significant support, love, and cheerleading from my immediate family—my mother, Iona; my sister, Lisa, and brother-in-law, Ian; my brother, Jay, and sister-in-law, Stephanie; and my fantastic niece and nephews, Juwan, Justine, and Ian Jr.—as well as from my extended family of aunts, uncles, cousins, neighbors, church family members, and friends. Thank you all.

ONE

INTRODUCTION

Who are Black mathematicians? What are their paths to the profession? Although this book provides some answers to these questions, it is by necessity a synthesis of many stories past and present. By one estimate, there are roughly 300 living Black mathematicians in the United States. They work in colleges and universities; for federal, state, and local governments and agencies; in private and public secondary schools; and in industry. Their fields encompass pure and applied mathematics, including operations research, analysis, game theory, topology, algebra, number theory, and statistics. The mathematics they do is elegant, relevant, and practical, as well as critical, for the sciences, technology, engineering, finance, public policy, national security, and a host of other domains relevant to the well-being of the country and the world.

Black mathematicians are indeed a rarity, as are mathematicians in the United States generally. Recent American Mathematical Society data show that roughly 1,400 people were granted the PhD in a mathematical science by a U.S. university in 2008–2009, the most recent year for which data are available. Half of those individuals hailed from outside of the United States. Of the 669 American citizens who received the PhD, 86 were Black, Latino, Asian, or Native American. Nineteen, or less than 3% that year, were Black.

In the public imagination, mathematicians seem to spring fully formed as individuals whose sole interest is mathematics, who are socially inept, and who are unconcerned with any topic other than mathematics. There is a prevalent idea that mathematics is a completely solitary enterprise, done in the absence of any community. Burton (1999) describes "a false social stereotype, promoted and reinforced by the media, of the (male) mathematician, locked away in an attic room, scribbling on his [*sic*] whiteboard and, possibly, solving Fermat's Last Theorem" (p. 127). These notions of mathematics and those who do it are disseminated to American school students at an early age. Substantial research in mathematics education reveals that both elementary and secondary teachers and students share limited notions of mathematics and, further, narrow ideas about who mathematicians are and the work that they do (Cirillo & Herbel-Eisenmann, 2011; Moreau, Mendick, & Epstein, 2009).

Works focusing on the formative and educational experiences of mathematicians are relatively rare. Within these works, Black mathematicians are often absent altogether or represented by one or two individuals. For example, books exploring women mathematicians (Murray, 2000) or scientists in general (Hermanowicz, 1998) do not usually include more than one or two African Americans. Books addressing Black mathematicians and their research for the most part do not focus on their personal and professional lives nor on the journeys they took to become mathematicians (Dean, 1997; Dean, McZeal, & Williams, 1999; Newell et al., 1980). Books that provide more detailed personal information and include Black mathematicians (e.g., Albers & Alexanderson, 1985; Albers, Alexanderson, & Reid, 1990; Kenschaft, 2005) do so largely in the style of encyclopedias, with brief synopses of Black mathematicians' personal and professional lives; Kessler, Kidd, Kidd, and Morin's (1996) profiles of 100 Black scientists include 4 mathematicians. *Beyond Banneker* provides detail about Black mathematicians' early mathematical experiences and in-depth analysis of the relationships between their cultural, ethnic, and mathematical identities. In documenting, describing, and analyzing the formative, educational, and professional experiences of Black mathematicians, this book seeks to add a significant missing component to the national narrative about mathematics and mathematicians.

Throughout this book, I primarily draw on extensive interviews conducted with 35 U.S.-born Black mathematicians who earned their PhDs in a mathematical science between 1941 and 2008 and describe the richness and variety of Black mathematicians' experiences, from Banneker and Fuller's time to the present day. The interviews are augmented by data collected from books, texts, oral histories, articles, and essays written by and about Black mathematicians in the United States and by information gleaned from site visits to conferences, colleges, and universities.[1]

I begin by telling the stories of the first Black persons in the United States to earn their PhDs in mathematics, Elbert Frank Cox and Euphemia Lofton Haynes, and of the three oldest Black mathematicians in the United States interviewed for this study, David Blackwell, Evelyn Granville, and Clarence Stephens. These mathematicians, the "Vanguard," reveal that a host of factors contributed to their becoming mathematicians. As they were among the first Black Americans to earn their PhDs in mathematics, their stories have inspired others to pursue mathematics.

The Vanguard

> When I was growing up, it [was] highly unlikely that you would know a PhD. In fact, when I got to [this job] my good [White] buddy would say, "Wayne, when I was growing up I had an uncle who was a mathematician so I kind of knew what they did and I knew that I wanted to be one. But in your case, how did you know you wanted to be one?" *Now, in my case it was different: you could not see all the way to a PhD.*
>
> —Wayne Leverett, PhD, interview[2] (my italics)

In any category of "firsts," there is always an underlying question of how "the first" decided that it was within his or her power to become that which no one like him or her had become before. The answers Elbert Frank Cox and Euphemia Lofton Haynes (the first Black male and female Americans to earn doctoral degrees in mathematics) might have given to this question, unfortunately, are unknown. But what we do know about them—largely through remembrances from Cox's students, colleagues, and family members and from Lofton Haynes's collection of family and professional papers held at Catholic University—tells us something about their journeys as mathematicians and, also, the ways in which these journeys left blueprints for others to follow.

Elbert Frank Cox, who earned his PhD in mathematics from Cornell in 1925, was born in Evansville, Indiana, in 1895. Indiana's location in the Midwest, bordering southern states but also serving as a home for abolitionists, meant that it had attendant characteristics of both the North and the South with regard to race relations. Cox lived on a neighborhood block that was considered to be racially mixed, but in 1903, the "most serious race riot in Evansville history broke out in his neighborhood" (Donaldson & Fleming, 2000, p. 106). Yet despite growing up in a racially mixed neighborhood, Cox attended segregated schools for most of his childhood. His father, Johnson D. Cox, was a teacher at the elementary school he attended. Cox's accomplishments are all the more impressive given that he was born and came of age during a period considered to be the nadir of American race relations—a period of curtailed and withdrawn civil rights and extensive racial violence against Blacks between the end of Reconstruction and the beginning of the new century (Anderson, 1988).

Cox enrolled in Indiana University in 1913 as a mathematics and physics major, earning all A's in his mathematics courses. After graduating from there in 1917, he became a high school mathematics and physics teacher in Kentucky. He then began teaching at Shaw University in Raleigh, North Carolina, in 1919. While teaching at Shaw, he began taking courses at Cornell University, beginning in 1920. He applied to Cornell's doctoral program in 1921 and solicited letters of recommendation from his Indiana University professors. One professor, S. C. Davisson, wrote that he "would be glad to have him in the graduate program in Indiana: 'he surpasses any colored man I have known as a student in mathematics'" (Donaldson & Fleming, 2000, p. 112). Another, probably Tobias Dantzig, wrote an official letter, and then wrote a second letter to Professor Tanner at Cornell because he anticipated "certain difficulties for the young man because of the fact that he is of the colored race" (Donaldson & Fleming, 2000, p. 112).

Kenschaft (1987) notes that Cox's Cornell advisor, William Lloyd Garrison Williams, was probably aware of the significance of Cox being the first Black person in the world to earn his PhD in mathematics and suggested that he submit his dissertation abroad in addition to Cornell so his "status could not be disputed" (p. 172). Cox finished the PhD in 1925, and after a few years as a professor at West Virginia State College, he began teaching at Howard University, in 1929. At Howard, he joined Dudley Woodard (PhD 1928), the second African American to earn his PhD in mathematics (from the University of Pennsylvania), who was the department chairman. Eventually, the mathematics department at Howard could claim as faculty five of the first eight African Americans to earn their PhDs in mathematics, including William Claytor (PhD 1933, University of Pennsylvania), David Blackwell (PhD 1941, Illinois), and J. Ernest Wilkins (PhD 1942, Chicago). Although Howard did not have a PhD program in mathematics until 1975, its master's degree program in mathematics was one of the first among historically Black colleges and universities (HBCUs) and produced several students who eventually went on to earn their PhDs. Upon Cox's retirement in the 1965–1966 school year, Howard's president remarked that Cox "had directed more masters degree students than any other professor at Howard" (Donaldson & Fleming, 2000, p. 121).

Since his retirement and death (in 1969), Cox's life and career in mathematics have been honored, most prominently with the Cox-Talbot Address by the National Association of Mathematicians (NAM), an organization founded by Black mathematicians. Whether his path crossed that of a fellow Washingtonian who "in her spare time taught at Howard University," Euphemia Lofton Haynes, the first African American woman to earn her PhD in mathematics, is unknown.

Lofton Haynes's accomplishment itself as the first African American woman to earn her PhD in mathematics was unknown for many years, and in fact, Marjorie Browne (PhD 1950, Michigan) and Evelyn Boyd Granville (PhD 1949, Yale) were each assumed at various periods to be the first African American women to earn their PhDs in mathematics.[3] Despite the fact that some reports published around the time of Lofton Haynes's death mention her receiving the doctorate in mathematics from Catholic University in 1943, no one in the mathematics community seems to have connected the dots until the late 1990s. Granville, born and raised in Washington, DC, herself an alumna of the same secondary school that Lofton Haynes attended, noted, "surprisingly, no one in DC ever mentioned the name of Euphemia Lofton Haynes to me and I did not hear about her until late 1999. This remains a mystery to me" (E. B. Granville, personal communication).

Born in 1890, Lofton Haynes was a contemporary of Cox, and although she lived in Washington, DC, for much of her life—the exceptions being her attendance at Smith College (bachelor's degree in mathematics, 1914) and the University of Chicago (master's in education, 1920)—it seems that their paths never crossed. Before leaving DC for Smith and the University of Chicago, Lofton Haynes began her education in the segregated city schools of Washington, DC, attending the highly regarded Dunbar School[4] and graduating from it as valedictorian (Duffie, 2003).

What made Euphemia Lofton Haynes decide to pursue her doctorate in mathematics is unknown, but her commitment to education was clear. Teaching for 47 years in the DC public school system and also as a professor at Miner Teachers College in DC (now part of the University of the District of Columbia), she became a member and, eventually, president of the DC Board of Education. Widely acknowledged as a key factor in the integration of the DC schools in the 1960s and 1970s, she was a fierce opponent of tracking in schools. When she died in 1980, her status as the first Black woman to earn a PhD in mathematics was unknown, despite her leaving a substantial collection of family papers to Catholic University.

Many of the teachers at Dunbar, as Evelyn Boyd Granville attests and other chroniclers have described (Cromwell, 2006; Sowell, 1974), were highly educated and influenced their students to pursue postsecondary education. Lofton Haynes fondly remembered Miss Harriette Shadd, a teacher at

Dunbar and a Smith College graduate—"I just idolized her, that's all" (Lofton Haynes oral history). It was due to her influence that Lofton Haynes wished to attend, and eventually enrolled in, Smith. All we know of Lofton Haynes's decision to get her PhD in mathematics is what she told an interviewer in 1972[5]:

> [I] approach everything from a philosophical point of view. Does that say anything? I have been a mathematics scholar all of my life, through high school, through college, and then to get my doctor's degree in mathematics. Now I didn't expect to get my doctors degree, never, in mathematics but I wasn't surprised . . . because I enjoyed it so much.

Three of the oldest mathematicians interviewed for this book—David Blackwell (1919–2010), Evelyn Boyd Granville (1924–), and Clarence Stephens (1917–)—share some similarities with Cox and Lofton Haynes (including the fact that all of them have ties to the greater Washington, DC, area), although the three grew up in very different circumstances. Blackwell (arguably the most well-known) was born and educated in a predominantly White small town in Illinois; Granville was born and raised in predominantly Black Washington, DC; and Stephens was born and educated in segregated schools in rural and urban North Carolina.

Despite these different backgrounds, these three mathematicians share a common experience: they were Black and earned their PhDs in mathematics in an era when to be Black and highly educated was quite rare. David Blackwell was the seventh Black person to earn his PhD in mathematics, and he received it in 1941 from the University of Illinois, Urbana-Champaign. He was the first African American to be inducted into the National Academy of Sciences in any field, and he retired in 1988 from a long career as a professor of mathematics and statistics at the University of California at Berkeley. Despite his growing up in Centralia, Illinois—a predominantly White town that he noted was "*not* North of the Mason-Dixon Line"—Blackwell felt that he "really didn't face any obstacles to becoming a mathematician" (D. Blackwell, personal communication, October 2007). For example, although he attended a predominantly White high school with an all-White teaching staff, one of his teachers recognized his mathematical talent and encouraged him to join the mathematics club.

Blackwell did not suggest, however, that his induction into the profession was a completely color-blind one. He described this experience: after he completed his undergraduate degree, he was being considered for graduate fellowships, along with other White candidates (Agwu, Smith, & Barry, 2003). One of the fellowships involved teaching; the other, with greater funding, was a nonteaching fellowship. One of his White peers told him that he would probably get the nonteaching fellowship, saying

> "Well, you're good enough to be supported one way or another. And they're not going to put you in front of a classroom" . . . and of course, he meant because I was Black. And you know, he was right? (D. Blackwell, personal communication, October 2007)

Evelyn Boyd Granville (PhD, 1949) was one of the first African American women to receive a PhD in mathematics. Born in Washington, DC, she attended segregated schools, including the aforementioned Dunbar High School, which has been renowned for its history in educating Blacks in Washington, DC, for decades (Sowell, 1974):

> Dunbar gave us inspiration, quality education, and, you know, they made us feel good about ourselves. So they gave us, I can't think of a good word, but self-something-or-other. . . . It was a tradition at Dunbar to encourage us to go to the Ivy League schools. And Miss Mary Cromwell [one of Granville's mathematics teachers] was the sister of Dr. Otelia Cromwell, a graduate of Smith in 1900, somewhere around there. Otelia Cromwell went to Smith, and then later went to Yale and got her PhD in English. And Miss Mary Cromwell and Dr. Otelia's niece [Adelaide][6] also went to Smith. They encouraged me to apply to Smith, but I also applied to Mount Holyoke. And I was admitted to both Smith and Mount Holyoke, but I chose Smith, I'm sure at the urging of the Cromwells. (E. B. Granville, interview, 2009)

Granville completed her doctorate in mathematics at Yale, then spent some time in New York City doing postdoctoral work. Eventually, she accepted a position at the historically Black Fisk University in 1950, where she taught Etta Zuber Falconer (PhD 1969, Emory University) and Vivienne Malone Mayes (PhD 1966, Texas)—who once commented that it was the presence of Granville that influenced her and others to pursue the PhD. Then Granville began a career at IBM before returning to academe and retiring in Texas. Asked once to summarize her accomplishments, Granville stated: "'first of all, showing that women can do mathematics.' Then she added, 'Being an African-American woman, letting people know we have brains too'" (Young, 1998, p. 212).

Clarence Stephens (PhD 1943), like Blackwell and Granville, has been a faculty member at both HBCUs and predominantly White institutions. Born in rural North Carolina in 1917, he attended segregated elementary schools and an all-Black boarding school, the Harbison Institute, for secondary school. Graduating from Johnson C. Smith College, a historically Black college, he earned a master's degree and a doctorate from the University of Michigan in 1941. He is the ninth African American to earn his PhD in mathematics. Starting at Prairie View University, Stephens then began an illustrious career at Morgan State College (now Morgan State University) in Baltimore, Maryland, in 1947, followed by a well-documented and equally successful career at State University of New York (SUNY) Potsdam (Datta, 1993; Megginson, 2003). Early in his career, he was committed to increasing the number of mathematics majors at Morgan State and began recruiting students from the segregated Black high schools in Baltimore:

> There were two high schools in Baltimore: Dunbar and Douglass. . . . I went to Dunbar first. When I talked to the teachers they recommended [a top student], but they warned me that if we instilled ambition in all of the students they would be frustrated because they couldn't get a job [due to job discrimination against Blacks]. They were telling me about some person who got an engineering degree. Once he got it, he couldn't get a job. I said, "Nonsense," . . . I never pa[id] much attention to that because I just feel that if you get a good education, you've got a good education. (C. Stephens, interview, 2009)

Despite low expectations for student success from some of his Morgan State colleagues and administrators, Stephens was successful at not only increasing the number of mathematics majors but also at facilitating students' entry into PhD programs. Leaving Morgan State in 1962, he continued his model program at two campuses of the State University of New York (Geneseo and Potsdam) and has received numerous commendations for his mentoring and expertise in attracting college students to mathematics. By his reckoning, Stephens had about 10 students from Morgan—a small college in segregated Baltimore—in a 15-year period who earned a PhD in mathematics—and notably, Stephens's program had at least three students from a single class year at Morgan State who earned their doctorates in mathematics: Earl Barnes (PhD 1968), Arthur Grainger (PhD 1975), and Scott Williams (PhD 1969). As Arthur Grainger recalled:

> Earl, Scott, and I met the very first day of orientation here at Morgan. We had all determined that we knew our major was going to be in mathematics.

Grainger vividly describes Stephens "setting the stage" for developing mathematics students:

> [At Morgan State, Stephens] made the effort to try and get as many talented people in math together. We were in one of these calculus courses just for math majors. He came in the first day and he had a picture that he put on the bulletin board, a picture of a nice building. It was the Institute for Advanced Study at Princeton. That was the picture that he put up. And then he told us, "You are to aim for here; to get here at the Princeton Institute." You know, where Einstein was. He said, "Even if you miss, at a minimum, you will have a PhD. Because to get there, you would have to have a PhD, so if you aim here, you can miss." So in other words, establish that and set your goals very high in aiming at that so that if you don't make it, you will be pretty well off. (A. Grainger, interview, 2008)

These short introductions to the first Black mathematics PhDs in the United States (including two living elders, Granville and Stephens) reveal that many factors, including their mathematical talents and interests; thwarted potential and the awareness of limited opportunity; the sense of being in the right place at the right time; the supportive and rigorous educational environment at their schools, whether segregated or not; and the importance of family and community networks, all contributed to their becoming mathematicians. These experiences shaped and formed these mathematicians and, in turn, inspired others to pursue mathematics. Their stories are largely overlooked reflections of American history, mathematics, and education across the 20th and 21st centuries. In many ways, their mathematical lives are metaphors for the Black experience in America—opportunities earned, granted, denied, rescinded; civil rights as citizens upheld as well as challenged; and an ever-present "double consciousness" of what it means to be Black in educational and professional settings.

One Mathematician's Story

To illustrate some of what is overlooked in discussions about mathematical excellence and how it is attained, here I present in detail one Black mathematician's rich narrative about his mathematical experiences. Wayne Leverett, who earned his PhD in the 1960s, represents roughly the midpoint of Black mathematicians attaining their PhDs: squarely between Cox and other early mathematicians' achievements in the 1920s and 1930s and those of the youngest mathematicians in this study, who earned their PhDs in the 1990s and 2000s. The structure of his narrative, as I present it—beginning with key mathematical memories and the role of family, community, and school in fostering mathematics learning, segueing into professional and induction experiences in the field—is the same organizing structure that is used for the rest of this book. Throughout Leverett's narrative—as well as throughout the rest of this book—there is an understanding that history plays a part in his mathematical journey. His narrative about his mathematics experiences describes how he became a mathematician and, importantly, demonstrates how community and historical contexts are integral to the development of one's identity and success. Within the narrative are the pervasive twins of obstacles and opportunities, as well as more than a few occasions of serendipity.

Mathematical Memories

Leverett recounts his first mathematical memory:

> One thing I remember is that when I was in about ninth grade, my uncle worked for a construction company. He saw the foreman using a slide rule. He just got curious about it, so the foreman said, "Well next time I place an order for equipment, I will order you one if you'd like." So the slide rule came with a thick manual about trigonometric functions and those such things. It was way over my uncle's head. He was a carpenter, trained to do carpentry on the GI Bill. In the family, people thought that I was some sort of bookworm because I was always reading books. So he just gave it to me.
>
> I wanted to get to the basics of the thing. I wanted to understand it, so I actually read the manual. I knew enough algebra and trigonometry to figure out most of the scales. For me it became a

> hobby. So at school when the teachers discovered that I could use this thing, they were quite amazed. . . . This slide rule was one of my first memories about experiences that got me hooked on math for sure.

Leverett obviously had some school mathematics memories before this one—"I knew enough algebra and trigonometry to figure out most of the scales"—but this is, for him, a central formative mathematics experience. Why? Embedded in this story is his family's understanding that Leverett was something special—a bookworm who would appreciate the gift of a book, even if it was only a manual about trigonometric functions. Later in his narrative, it becomes clear that Leverett sees this moment as being critical to his pursuit of mathematics and development as a mathematics learner.

A related story that figures prominently in Leverett's mathematical memory is the experience he and a friend had with "mathematics in action."

> When I was supposed to graduate [from high school], a good buddy of mine and I were idling time away walking down a country road headed home, I believe. We came upon a little White man who was surveying some land. He needed two strong fellows to help him pull some chains. He told us that he would pay us 75 cents per hour to do this. This is a lot more than you could make working on the farm. You could earn two or three dollars a day by working on the farm, but here is a guy who is going to pay 75 cents per hour. I thought that this was an enormous sum of money to pull these chains, however, when this guy starting talking to us, he had a transit. He would set it up in sight through here and swing around through a certain angle and sight through there. He could compute the distance between two far-away points. When he found out that I knew a little trigonometry, he started teaching me how to use this transit. He was so impressed with me and I was so amazed by how much money you could make using this trigonometry. So I said right away that I wanted to be an engineer because I thought that engineers made even more money than high school math teachers. I wanted to be a civil engineer.
>
> This was a moving experience. I wish that students at tenth grade level could see something like this where here is something I am learning in school that is being used to earn money. Meeting that engineer who was surveying land. . . . I am glad I remember that story because of the fact that this guy could give $1.50 from his

> salary. Based on salaries that I knew about, from farmers, this guy must be making a lot of money. And he was friendly enough to teach me things about how he was actually measuring the distance in doing this without having to jump across that ditch over there to get to here. Now we had studied about triangles and all—if you know this side and you know this side and you know the angle between you can get the length of the third side and all that. But here it was in action. This was very powerful. So by the time I graduated from [college], I had not forgotten about being an engineer, all of that money.

Although the preceding stories tell us something about Leverett's out of school experiences with mathematics, it is clear that in-school and out-of-school mathematics reinforced each other in powerful ways:

> I remember, maybe in the tenth grade algebra class, she [the teacher] gave me half the class [to teach]. My first memory of doing math [in school] was as a show-off. I was having fun, but I think the fact that the teachers gave me praise really encouraged me to do a bit more. When we were taking algebra, Mrs. Barr gave me a college algebra book because I think she feared that I could keep up with the regular algebra easily. She gave me a college algebra book and would check off a couple of problems and say, "See if you can do these tonight." I would go home determined to do them because I wanted to stay in her good graces. She thought I was smarter than I was and I wanted to keep it that way. So I would work on the problems, sometimes, half the night before I would figure out how to solve them, but I would come in the next day as if I had solved them in 15 minutes. "Here is the solution, give me some more." I managed to keep that going until I graduated. . . . [A]t the end of the year when I tried to return the book, she said, "Wayne, you keep that book. It will do you more good than it will do me." I thought it was such a great treasure to have her book.
>
> The only thing that I have taken to everywhere I go is to remember what teachers did for me when I was in high school. Because if Burgess had ignored me, or if Barr had ignored me, or [his college professor], I don't know where I would be today. I certainly wouldn't be here. So when I see a student who has some ability and is trying, I always try to pull them aside and do something special. I keep looking for students to befriend and yes, I try to find a good student to mentor and watch them and see how they grow.

Throughout Leverett's narrative about his early mathematical journey, he makes connections to his own current practice as a mathematician and his own philosophy about mathematics. He highlights the importance of young people understanding the real-world implications of the mathematics they learn in school and also seeks to emulate his own high school teachers and college professor in his work as a mathematician at a research university.

Socialization Experiences within the Field

One of the most telling quotations from Leverett's narrative is a question from a colleague who was working with him in private industry: "Wayne, when I was growing up I had an uncle who was a mathematician so I kind of knew what they did and I knew that I wanted to be one. But in your case, how did you know you wanted to be one?"

The assumption of Leverett's colleague that Leverett would not have known "how he wanted to be a mathematician" in the traditional way—via a close network or family member—was not an unfounded one, given the era. In addition, what we know about Leverett's family from his own story suggests that most were not college educated, and this underscores that the traditional routes to mathematics careers might not have been available to Leverett.

But it would be dismissive—and erroneous—to discount the critical contributions of Leverett's family members (who had not received the same opportunities for education as those of his colleague's family) to his development as a mathematician. In particular, two of Leverett's uncles, in addition to his father, figure prominently in his narrative:

> My uncle was a unique person. He was always curious about everything: art, technology. The rest of my family, of course, always supported and encouraged me to be a good student, but they didn't know technology, and they couldn't help me with algebra. He couldn't help me, but at least he was brave enough and curious enough to be interested in the slide rule.
>
> My parents had been divorced and my father moved to [another city] and worked for a steel company. My dad had a brother in that city, Uncle Fred. By then they all assumed that I was a pretty good student. They all thought that I was smarter than I was. I really thought that I was pulling a great big scam on all of them, but I never told them that. So one day I am [visiting and] riding around

> with my uncle and he says, "Wayne, you say you want to go to college and blah blah blah." I said, "Yes. I would like that very much." He says, "We have a college here; let's try it out tomorrow and take a look at them."
>
> I remember he was quite a brave guy because we went to the registrar's office and there was some person there like a receptionist or secretary. She tried to give us some application papers to fill out. My uncle said, "We don't want to fill out papers, we want to talk to the registrar." They actually brought him out and we spoke to him. Uncle Fred told him how smart I was in his opinion. [The registrar] basically said, "Look if he is all you say he is, we will admit him. Have him send his high school transcript when he gets back home. It is going to cost $350.00 a semester or year." I can't remember, but I remember the $350.00. So I basically signed up there on the spot just pending my high school transcript being as good as my uncle had claimed it was. So I went home and told Dad that I had signed up for college that day. The good thing is this: they said, "If your father has been living here all these years, you could be admitted as a resident of this state." So there is another lucky break, that some uncle who really thought I was smart decided to take me over and at least investigate this college.
>
> My Dad worked there [at the steel mill]. It was a very strong union. It was amazing. So the pay was good for unskilled labor. But he used to say to me in college, "Wayne, I hope you never have to work at this company." Actually, there was a lot of pollution and dirt and smog. I could tell by the black stuff that would be on his clothes from burning coal, which is something that they did to cook the steel. But yes, he made quite enough money to pay my tuition. So I went through college without having to work.

This network of African American men, none of whom had gone to college, supported Leverett's development in multiple ways. The first uncle arguably set Leverett on the path to wanting to learn as much as he could about mathematics. The second uncle single-handedly (and later backed up by Leverett's strong high school record) got him enrolled in a Black college that happened to have a strong emphasis on mathematics. And Leverett's father had the vision of wanting more for his son than a good-paying union job and supported him financially throughout college so that he could focus on his studies.

As Leverett reveals, at every stage he saw more and more opportunities for his life and career:

> Well, I mean, when I was in high school and in [the rural South] in farming country, and segregation, and hardly any industry anyway, Black professionals tended to be preachers or teachers. Everybody else was a farmer. So I thought that being a teacher was a good occupation that provided a fairly good lifestyle and that is what I wanted.
>
> Now I didn't know anything about PhDs. I didn't know any PhD. My goal was to earn a BS degree in mathematics and become a high school math teacher. But when I got to [college], there was Dr. Stephens, who has won all kinds of awards for his teaching. He really stretched us and he started talking to us about the PhD. When I got to [college], I had changed my plans. I said, "Well, you know, I think I would rather teach at [this college] than to teach at East Side High." So I kind of started to think about teaching in college rather than teaching at the high school level. [When I realized] that the college professors get to talk about more interesting math at a higher level and how you learn more beautiful things and work with smart students, then of course, I changed my goals.

What Leverett likens to a contemporary research undergraduate program in mathematics (research experiences for undergraduates, or REU) was really his induction into a community of mathematics doers. He met other talented mathematics students, and all of them were supported by a talented professor of mathematics. Stephens's program, in addition to Leverett's father's financial contributions, ensured that his focus could be on his mathematics work and on excelling in mathematics. His excellent grades and performance on the Graduate Record Examination (GRE) ensured Leverett's fellowship and his smooth path through graduate school on the way to the PhD.

Opportunities Then and Now

As Leverett stated earlier, in his case, "it was different: you could not see all the way to a PhD":

> You know, it's different these days. Kids have the internet. They can look up a whole bunch of information about [careers]. And it's

> scary, I mean, I think back how it happened, I went [to the company] without knowing very much about [what the company did]. I mean, I didn't know what they made, what kind of problems they were trying to solve. It was a lucky stroke, which I don't think happens today. I think they hire mathematicians but they're much more project-oriented. They will hire somebody who can come in on day one and help out with this signal processing thing they're interested in and they don't just look at somebody's transcript and say, "Come on in, you got good grades, we're going to try to get you interested in some problems we have and hopefully you can help us out." They don't do that anymore. But I'm sure any old person you talk to will say that things are much different today than when I grew up and that's all I'm saying. One has to be envious of the opportunities that young kids have these days.

Wayne Leverett's narrative shows the influence of his earliest memories of mathematics: they are rooted in the recognition that his father and uncles helped hew a path for him to become a mathematician, and that serendipitous encounters—with a surveyor, with a professor, and with like-minded students at his undergraduate HBCU—helped him see that mathematics could be a career goal. Although much of Leverett's mathematical journey follows the traditional path for professional mathematicians—college attendance and apprenticeship with a mathematician and graduate school immediately following—his story reveals much more about how people might find their way into *becoming* mathematicians. These stories, although largely missing for Elbert Frank Cox and Euphemia Lofton Haynes in their own words, are critical not only to understanding how more of our talented citizens might fully reach their potential but also to illuminating a field that privileges itself as largely unknowable to most. The goal of this book is to share these stories, in an effort to make mathematical lives known and knowable. The remaining chapters of this book show the paths, spaces, and communities that contributed to these mathematicians becoming who they are.

Organization of Remaining Chapters

Beyond Banneker describes the formative, educational, and professional experiences of 35 mathematicians interviewed between 2007 and 2010. Throughout the remaining chapters, I draw on the experiences of mathematicians I interviewed. To provide some context vis à vis the time periods that mathematicians "came of age," I have divided the 35 into three "generations." The first generation includes three mathematicians who are referenced without pseudonyms (David Blackwell, Evelyn Granville, and Clarence Stephens) throughout this book. First-generation mathematicians earned their PhDs before 1965, largely before the gains of the Civil Rights Movement. There are 12 second-generation mathematicians who earned their PhDs between 1965 and 1985, four of whom earned their PhDs before 1970. Many second-generation mathematicians attended high school in the 1950s and 1960s, at the zenith of the Civil Rights Movement. Finally, there are 20 third-generation mathematicians who earned their PhDs between 1985 and the present; four of them earned their PhDs before 1990. Most of these mathematicians were high school students in the 1970s, 1980s, and 1990s.

The remaining chapters follow the "mathematical life spans" of mathematicians from early interest and educational experiences in mathematics to their professional lives as mathematicians. Chapter 2: Kinships and Communities describes how most mathematicians' first mathematical memories are rooted in family experiences and critical school experiences in childhood and adolescence. It also lays the foundation for descriptions of how fictive kinship networks emerge for Black mathematicians within schools, professional networks, and the field. Chapter 3: Navigating the Mason-Dixon Divide, focuses on "border states" of being: the chapter title takes its cue from David Blackwell's statement about his hometown of Centralia, Illinois, being "*not* North of the Mason-Dixon Line." In particular, it examines the role of the space and place of the South in facilitating mathematics learning and development in caring all-Black schools and colleges and also posing challenges to opportunities for learning. These experiences cut across generations of mathematicians and are shared by some of the youngest participants in the study as well as some of the oldest. This chapter also focuses on the mathematical and educational experiences of those who were among the first to desegregate all-White high schools and colleges, including a number of second-generation mathematicians who did so in the 1960s. Chapter 4: Representing the Race takes its title from a 10th grader who once commented to me that being Black in a math class and getting called on meant that one was representing the entire race when answering a math question. Being wrong in this setting would have had dire consequences in terms of how Black students were viewed. This chapter revisits the histories of Thomas Fuller and Benjamin Banneker and describes Black mathematicians' narratives of their induction into the profession as adults, focusing on their experiences in graduate school and their professional careers. Further, it explores how they see themselves represented by others and how they define themselves. It examines how these representations present opportunities for and challenges to their mathematical performance and participation. In addition, this chapter discusses the place of Black women mathematicians, whose race and gender operate together to present unique opportunities for and obstacles to their mathematical development. Chapter 5: Flying Home describes the importance of historically Black colleges and universities in cultivating mathematics talent, the nostalgia held by mathematicians (those who were and who were not educated at HBCUs) for them, and the mission of some mathematicians to return there. Further, it describes the efforts of Black mathematicians at predominantly White institutions to create structures and communities that in some ways mimic those of HBCUs. This chapter builds on the discussion of the networks initially described in chapter 2. Chapter 6 is the conclusion, which revisits the lives and experiences of the Vanguard in light of what we have learned from the lives and experiences of contemporary Black mathematicians in the United States. It revisits the paths to the doctorate and the profession of mathematics, the challenges that still face Black mathematicians, and the opportunities that they have seized for themselves and created for others. It describes emerging efforts of mathematicians to reach younger and younger generations of Black students and suggests additional avenues for exploration and research.[7]

TWO

KINSHIPS AND COMMUNITIES

The mathematical journeys of Elbert F. Cox and Euphemia Lofton Haynes described in the introduction suggest a compelling question: how does someone who has never known a mathematician decide to become one? Wayne Leverett's self-described trajectory—exposed to the route of becoming a mathematician in stages, first in high school through his teachers, then in college through his mentor and major professor, and so forth—is a story shared by many first- and second-generation mathematicians, whose experiences growing up in the first half of the 20th century starkly demonstrate just how narrowly opportunities were defined for African Americans during that time period. As Evelyn Boyd Granville, one of the first Black women in the United States to earn her PhD in mathematics, recalls:

> I think my sister was good in math. But I don't remember exactly; I know she started college. But she didn't finish, she stayed one year and then she didn't finish. I don't know whether she was going to be a math major or not. I think she was pretty good. Now, my mother, I don't recall my mother being particularly adept at mathematics. And of course my mother didn't go to college, my father didn't go to college, so I don't know what their real strengths would have been had they been living in this current situation, in today.

Without close examination, it would be easy to come away from some first- and second-generation mathematicians' stories with an erroneous assumption: that because of the usually limited formal education of their parents and families and the lack of state and local government funding bestowed upon Black public schools at that time, Black mathematicians' social networks outside of school and educational experiences within school did not provide meaningful mathematics opportunities. The lives of many mathematicians across generations—with their rich descriptions of mathematics teaching and learning, both within and outside of schools—suggest differently. Even within the most challenging environments perceived as unlikely to provide opportunities for learning mathematics—one-room schoolhouses in the rural South, large high schools in urban centers—exemplary learning and teaching took place. And despite contemporary "postracial" rhetoric, stories of racial obstacles to younger mathematicians' development share some surprising similarities with those of older mathematicians.

Various networks and communities emerge as pivotal to Black mathematicians' development. These "kinships"—both fictive and familial—are formal and informal, planned and serendipitous. Fictive kinships are close, familylike relationships among individuals who do not share blood ties, and social scientists have found them particularly prevalent and influential in the African American community, dating from the era of slavery. For example, some Black mathematicians, as do some members of the larger community of mathematicians, speak of their "mathematical brothers" and "mathematical sisters," peers and mentors who have studied with them, mentored them, or collaborated with them throughout their professional careers. Contemporary professional kinships owe much to Black mathematicians' mid-20th century efforts to organize themselves when national mathematics organizations would not have them as fully participating members. In addition, historically Black colleges and universities provide an extensive network of alumni, students, faculty, and administrators who see producing Black professionals in the mathematical sciences as one of their core missions.

But we are getting ahead of ourselves. What early experiences are critical to the development of Black mathematicians? The first kinships that support mathematical development, not surprisingly, are rooted in the family. As Black mathematicians recall, families encourage mathematical interest, promote learning about mathematics concepts, and provide insulation against low expectations of Black students in formal school settings. These family networks are critical for mathematics learning and socialization, as they are for most mathematicians. As Black mathematicians enter schools, these kinships may expand to include peers, teachers, and administrators. And as Black mathematicians enter colleges and universities, both family and school kinships have a lasting impact and share characteristics with the professional kinships that Black mathematicians develop with each other. Thus, kinships in some form—with elements of caring, nurturing, instructing, socializing, mentoring, and inducting—permeate the lives of Black mathematicians from their early childhood years throughout their professional lives.

Carrying on the Family Tradition

Laverne Richardson and William Burris are two of the youngest mathematicians interviewed for this book, and like many others, they have heard stories about where their mathematical talents "come from":

> They just said that . . . [my grandfather] worked with a university, on some kind of railroad. They just said that he really always loved math and kind of was a whiz at it, but nothing formal. He had no type of formal training. They just indicated that he always had a natural ability when it came to mathematics. I never actually saw any of this displayed. (Laverne Richardson, third generation)

> I'm not really sure because my dad was more like a mechanic. But he was born in, like, 1920. So I don't think he actually got an opportunity, you know, to stay in school, because his father died when he was 14 so he had to take up the trade of being a mechanic at that age. But he had pretty good stories; he recalled breaking apart a transmission and engine, putting it back together when he was, like, 10. So I mean, maybe, I got it from him. He seemed to be pretty smart, but he never got to show that. Actually, he said he used to rebuild transmissions to where they run pretty fast. And he would do it for this company. But actually he should have been patenting some of that stuff, but I don't know. (William Burris, third generation)

Discussions about mathematicians' formative experiences often include descriptions of how family members with extensive years of formal education and/or who were active in the field of mathematics themselves contributed to the development of the mathematicians concerned. But for a significant number, family members who were supportive of their mathematics learning were not themselves in mathematics professions and, indeed, had not received many years of formal schooling. Of the 35 mathematicians interviewed, 9 have parents with degrees beyond the bachelor's degree, but there are also several who have parents who did not finish high school.

Undoubtedly this is true for some mathematicians of any background, that they were the first in their families to pursue advanced mathematics, but what is unique to Black mathematicians is that opportunities to learn were legally limited for many of their extended family members. Discriminatory practices, particularly in the South, kept many talented African Americans from attending school, going to college, or pursuing professional careers. What is probably also unique to African Americans is that these restrictions affect even the youngest mathematicians, in that some of their parents or grandparents were unable to pursue higher education and professional careers. For these mathematicians, limited information about mathematics and social capital related to how to pursue mathematics might have posed obstacles to their success in school. In many ways, the dichotomous experiences of Benjamin Banneker and Thomas Fuller are reflected in the family stories of Black mathematicians—both men exhibiting some degree of mathematical talent but experiencing widely varying opportunities to demonstrate that talent.

Like the stories recounted about Thomas Fuller's mathematical prowess, Black mathematicians' stories about forebears and their mathematics abilities often seem mythic. Richardson's, Burris's and others' narratives suggest that family narratives about forebears' mathematics experiences are powerful and resonant—but these word-of-mouth stories are often passed along second- and thirdhand, in that the mathematicians themselves may not have direct knowledge or have had conversations with the forebears about those mathematics experiences. Several mathematicians, across generations, mentioned these kinds of stories about relatives who exhibited mathematics interest and talent but never got the opportunity to develop their potential.

But mathematicians also hear directly from their relatives about their mathematics experiences. Parents, siblings, and other family members share their love of mathematics with them:

> I realized as I got older that that love of math came from my mother, because she loved math and she would tell me stories about how in high school, the teacher would give them these math problems and she and her best friend were the only ones that could do certain math problems. (Eleanor Gladwell, second generation)

Thus, the myths, memories, and family histories relating to mathematics recounted by mathematicians demonstrate the variety of methods with which family members can support mathematical development, in sometimes surprising ways. And for many mathematicians, certain members of the family were much more active than others in fostering mathematics success.

Although Wayne Leverett, unlike his colleague in private industry, may not have had an uncle who was a mathematician, his family members were deeply interested in developing his intellect. As described earlier, one uncle in particular was among the first to expose him to mathematical ideas. Leverett's description of his uncle as a kind of Renaissance man, interested in art and technology despite his having not graduated from high school, may seem unusual. But in an era when the very existence of secondary schools for African Americans in the South was a rarity, it is not at all unusual for first- and second-generation mathematicians to have family members (particularly parents, aunts, uncles, and grandparents) with only an elementary or junior high school education. And, indeed, even when secondary schools were available, the economic necessities of the time often prevented Blacks from attending—instead, they worked. Finally, for even the most highly educated African Americans, employment discrimination ensured that they were often limited to jobs as ministers or teachers in the Black community or, often, jobs as laborers or servants. The opportunities for highly educated Blacks were stratified and narrow. Despite these realities, education was viewed as critically important to better life chances, and these stories underscore that family members wanted these young mathematicians to have opportunities that they themselves might never have had.

Of the extended family members who mathematicians report as being influential to their mathematics success, uncles like Leverett's appear frequently in mathematicians' narratives. For example, David Blackwell has shared that his "Uncle Dave," who did not attend school beyond the elementary grades, was really good at "figuring"—adding columns of numbers quickly. This impressed the young Blackwell greatly. Several mathematicians also described mathematical relationships with uncles:

> Uncle Ryan was earning a master's degree and living with us. He would play memory games with me, counting large numbers and determining large numbers. Those were my earliest memories. My guess is it would have started when I was maybe four or so. Now after that, I just loved any math I got my hands on. (Ryan Kennedy, second generation)

> My uncle had always had aspirations—my uncle Bob—to be an engineer, and he would talk about mathematics a little bit, and [we'd do]

various little puzzles and that sort of thing—brain teasers. (Nathaniel Long, second generation)

Q: Were any other family members mathematically inclined?

A: Not really. I mean, not explicitly. On my mother's side there were a number of uncles and my grandfather, for example. I should say a number of grand uncles and my grandfather was very clever at checkers and things like that. I think they would have been, you know, given the opportunity, they may well have displayed some mathematical talent, but, you know. . . . (Rebecca Winter, second generation)

The focus on grandfathers and uncles in these mathematical memories probably reflects that for first- and second-generation mathematicians, men were more likely than women to be involved in careers as engineers, construction workers, and such. Uncles and other extended family members often exposed these mathematicians to mathematical thinking that went beyond standard grade school mathematics:

My grandfather lived right around the corner from here [the college where he is now a professor]. I remember he would always have these mental challenges that he would give me all the time. . . . I actually use one of them in particular [when I'm teaching]. We were on the front porch and he was asking me—if he walked halfway to the end of the porch, and then halfway again, and then halfway again, and so on, how many steps would it take him to reach the end of the porch? And so, I may have guessed five or something, I don't know. So then he actually proceeded to do it, halfway, and then halfway, and then halfway, but the idea was that he was converging—he didn't use the term convergence, but he never actually reached it—but he got closer and closer and closer, and of course he didn't say within epsilon. . . . But anyway, I have fun when I'm teaching about convergence to really tap into it at this early level. One just because I have fun telling the story—but also to give my students an idea of the sorts of things they can do with their students, because some of them may go on to become teachers, or just with their grandchildren one day, whatever the case may be. These are the sorts of things that can really bring high-level things in very early and just challenge the mind and make you think. (Stanley Parker, third generation)

This experience, vivid in Parker's memory, is shared with students in his college classes. In sharing it, Parker not only elucidates a mathematical concept that is often difficult for students to grasp, but he underscores the importance of passing along mathematical ideas, that mathematics doesn't just happen in school, and that his students can use accessible examples of mathematics when interacting with their own students or family members to illuminate complex mathematical concepts.

Other mathematicians describe mathematical memories that involve other immediate and extended family members. One third-generation mathematician lived with her grandparents, mother, aunt, and brother in a multigenerational household. She remembered fondly that "every night after dinner we would have to go to my grandfather's room for him to talk to us about our homework. He helped me learn my multiplication tables. He would always have the coolest things for us to do with academic stuff—not just math related, but all kinds of stuff." Another mathematician's father was an electrical engineer, who often enlisted her and her sisters' help with projects around the house: "That's how we learned fractions, that's how we learned to add, by working on these different projects with him."

While it is true for most first- and second-generation mathematicians that they were among the first in their families to attend college and graduate school, several second-generation mathematicians belong to families in which some family members were highly educated. For example, Ryan Kennedy's mother had earned her bachelor's and master's degrees in mathematics, and although he cannot recall talking with her about mathematics, he remembers that

> there were always math books around at home, and so I would borrow them as soon as I could understand them. And from somebody who had a master's degree, [there were] some serious math books there.

Kennedy recalls a seminal experience with his mother from his adolescence:

> When I was in seventh grade, my mother took me by MIT to show me about all the famous mathematicians that went there, so I always wanted to do that. Math was just, oh I just loved it all, you know, as much as I could.

But Kennedy's experience was relatively rare among mathematicians of the second generation, whose parents were largely born too early to reap the benefits of the Civil Rights Movement through extended educational opportunity. However, the educational attainment of Black mathematicians' parents, grandparents, and other extended family members varies widely. Some third-generation mathematicians—as with the first and second generations—were the first in their families to attend college, and some of the second generation and the youngest belong to families with highly educated members. One third-generation mathematician, Elizabeth Ricks, stated:

> I guess I could say that I come from a family of educators. We're all very interested in education. My grandfather got a master's in education from Harvard sometime in the early 1950s. My uncle, my dad's brother, majored in math. My dad's other brother is also a college professor in math, so he loves math. So Uncle Lawrence got a B.A. in math and then Uncle Langston got a master's in math. He teaches in [Texas]. So those are the family members in math. (Elizabeth Ricks, third generation)

As some of the narratives above describe, quite a few mathematicians have family members who held jobs and participated in careers that used mathematics—engineers, mechanics, teachers, and insurance agents, to name a few. Teachers and professors in mathematical sciences (mathematics, physics, statistics) are represented in the family narratives of Ryan Kennedy, Eleanor Gladwell, and others. One mathematician recalls going to summer school classes with her mother, who was a high school mathematics teacher, when she was a junior high school student, and she believes this was what sparked her interest in mathematics. Thus, despite the limits of the segregated era, some Black mathematicians across all generations were influenced by family members who were in a variety of mathematics-related careers, not unlike other mathematicians and scientists.

In addition to adults in families who facilitated mathematics learning and socialization, there were several instances in which Black mathematicians' siblings provided support for mathematical endeavors. One mathematician remembers her sisters quizzing her on her times tables, and another remembers playing and inventing pattern games with his brothers. In a more indirect way, one mathematician recalls his family's influence on his being interested in mathematics:

> Part of my inspiration did come from my family. We didn't sit down and do homework together and all of that, but because I had so

> many brothers and a sister who had scientific habits of the mind, it kind of just rubbed off on me. You looked at things in a different way. The expectations were different. You almost had this expectation that you were going to be an engineer, in our household. (Wade Trimble, third generation)

Mathematicians' rich family narratives reveal mathematical memories and experiences during childhood and adolescence that can be categorized in multiple ways. Many of these memories are focused on *procedures* and *memory tasks*, such as learning times tables or how to compute sums. Still others are categorized by their emphasis on learning *enhanced mathematics content*, whether it resembles school mathematics (in the case of the sisters learning fractions while assisting their father with construction projects around the house) or mathematics for enrichment (as in the case of Stanley Parker and the porch problem) or for projects. Other experiences are focused on *mathematics socialization*, that is, fostering the development of a mathematics identity and piquing mathematics interest. Some mathematicians had childhood experiences that underscored that mathematics was a useful tool to use in solving problems; others had experiences that underscored the aesthetic quality and elegance of mathematics.

For the most part, the experiences described above occurred outside of school, within family networks. But some of the most powerful and resonant experiences with mathematics for these mathematicians, not surprisingly, occurred within schools.

School Communities and Their Traditions

Wade Trimble, who described previously a sister and brothers with "scientific habits of mind," attributes those habits in part to the high school they all attended, noting

> All six of us are very good with mathematics, and I argue that from the standpoint that my oldest brother is a senior engineer with [a car company]. I'm the second oldest, and then the brother under me was an engineer. Then my sister, she teaches math at the grade school level. Then my next brother, who is very good at mathematics, he's into management now, though. Then my youngest brother, who was an engineer, is now a system administrator for [a municipality]. So

> mostly all of us have some sort of math or engineering degree. Of course that's also a product of us going to [a science and mathematics–themed high school]. . . . All of us went to that high school, and that high school played a big part in us becoming interested in math or engineering. That's what that school is all about: mathematics and engineering. (Wade Trimble)

Like Trimble, several mathematicians in this study attended well-known schools for mathematics, science, and engineering, including Stuyvesant and Bronx Science in New York City and Baltimore Polytechnic Institute and Baltimore City College in Baltimore. But most of the mathematicians attended high schools that were not necessarily focused on math or science. In particular, a number of the mathematicians—especially those who earned their PhDs before the 1970s—attended primarily segregated institutions. Unfortunately, the traditions and history of mathematics education in schools and colleges that served primarily African Americans is largely unknown. The dominant discourse about segregated, all-Black institutions has been that these institutions were less rigorous in their curricula than segregated, all-White institutions, despite some evidence to the contrary (Anderson, 1988; Siddle Walker, 1996; Sowell, 1974). In segregated Washington, DC, in 1899, for example, "in examinations given all high school students, the colored high school [M Street, which eventually became Dunbar] scored higher than either the Eastern or the Western high schools [which were White]" (Green, 1967, p. 137). In fact, some of these institutions were populated with administrators and teachers who had earned advanced degrees higher than those of administrators and teachers at White institutions (Siddle Walker, 1996).

As many have observed, the symbiotic nature of Black communities and schools was critical to the success of Black students. As one mathematician describes her own rural Alabama community:

> There are a lot of little towns, I think, small towns where they just really turn out a lot of good people, solid people with educational backgrounds. There are a lot of towns that value education. In Georgia, Moultrie—you'd be surprised what Moultrie, Georgia, has turned out. There're just towns that really have strong educational values, and they come from these people, teachers who have gone on to be leaders in the community, I think. So in those towns, you

> will see Black people who have been on the board of education for a long time, for example. [My community] was one of those towns. (Eleanor Gladwell, second generation)

For first- and second-generation mathematicians who attended segregated Black elementary and secondary institutions, school communities operated as kinship networks. Teachers, such as the Cromwells in Evelyn Granville's remembrances of Dunbar, took an active interest in the prospects of their students, and communities rallied together to provide funding for Black schools when state and local funding was lacking for much of the era of legalized segregation. One might expect this phenomenon of schools serving as kinship networks to be the experience of mathematicians who grew up and went to elementary and secondary schools before and during the Civil Rights Era, but it is also true of some younger mathematicians who attended all-Black institutions in the contemporary South.

The teachers in these institutions were instrumental in contributing to students' mathematics socialization and development. Wayne Leverett, whose mathematical narrative was described in part in the introduction, described two teachers at his South Carolina school whom he views as integral to his becoming a mathematician. After receiving the slide rule manual from his uncle, Leverett described how his learning about trigonometry from the manual intersected with his learning in school:

> I wanted to get to the basics of the thing. I wanted to understand it, so I actually read the manual. I knew enough algebra and trigonometry to figure out most of the scales. For me it became a hobby. So at school when the teachers discovered that I could use this thing, they were quite amazed. I remember maybe in the tenth grade algebra class, she started to teach a lot of trig to the class. She gave me half the class. My first memory of doing math was as a show-off. I was having fun, but I think the fact that the teachers gave me praise really encouraged me to do a bit more.
>
> [Those two teachers were] Mrs. Burgess and Mr. Barr. I mentioned this many years later when they invited me back to the school to talk. I do feel that I had some sort of special treatment that at least two teachers at a very small school noticed that I had some abilities and they did it on their own. They didn't get extra pay, but they were essentially giving me after-school tutoring. Nobody, even

the principal, knew that these things were going on. So I never have enough praise for those two teachers.

Eleanor Gladwell also attended segregated schools throughout her childhood and adolescence. She reports:

> And then when I went to high school, I had a high school math teacher who was a younger man that was recently out of college, so that means he had lots of energy and enthusiasm about mathematics. And that was about the time of Sputnik. So they had all these institutes around the country to try to increase interest in math, and they had a lot of teacher's institutes. And so he would go to those. But as a result, he was really excited about teaching math.

> I think it's because of him [the mathematics teacher] that I really excelled in math in high school. For example, when my class got to trigonometry, the county would not allow them to teach trigonometry because there were not enough students; you had to have enough for a big class. Well, he only had five, six, or eight. So he decided that we needed trigonometry to go to college. He agreed that if our parents would bring us back in the evenings, he would teach us trigonometry. And they did. That's how we learned, that's how we got our trig.

Thus, in addition to their families, Black mathematicians' elementary and secondary teachers have a profound influence on their development and access to rigorous mathematics. For mathematicians across all generations, their school experiences reveal that opportunities for mathematical thinking are not solely centered on solving problems in the context of doing homework, schoolwork, or studying for examinations. One third-generation mathematician describes a seminal moment in her school mathematics experience:

> In sixth grade I had a fabulous math teacher. First of all the fabulous math teacher was a Black woman, and it wasn't my first Black teacher. Even though I went to White high schools I had quite a few Black teachers through elementary school and middle school. But this teacher was just fantastic. Everybody loved her, everybody knew that she was just the best. And she started each class with a puzzle.

> There was always these puzzle things going on. . . . I'm getting teary eyed, my goodness. We just started puzzles.
>
> Q: And was that different from how you'd been doing math in school before?
>
> A: Yeah, it was just, get problems. I don't know how we were doing math before. I'm not sure I remembered how she did math. I just remember being excited about going to her class and thinking through these puzzles and I remember her giving a puzzle and feeling like I had no clue how to answer that. I had no clue how to even begin to answer it. So it was really exciting, maybe trying to figure it out, or hearing other people's solutions, even if I wasn't the one who figured it out, but it just sort of being, this sort of process, not this cut and dry "yes, no, you're right, you're wrong" kind of thing.
>
> Q: Was that sort of the first time that you realized you liked math or you were good in it?
>
> A: It certainly wasn't that I was conscious of it at the time. I think I don't remember ever having trouble with math, but when I look back on it, I feel like that gave me the first taste of sort of proving things or thinking through ideas in a little bit of a systematic way. As opposed to. . . . Because I think coming up through the years everything was just plug and chug. I'm not sure if that was all that interesting to me, all that plugging and chugging.

Several mathematicians, regardless of generation, spoke to the importance of teachers demonstrating pride in their work and having high expectations for them. A second-generation mathematician, who had attended a predominantly Black school in a northern city, described geometry as being "really great fun" in part because of a teacher who was very encouraging and showcased his work.

> He used to say, "Well, you don't have to take the test. I know you know it." Like that, you know? He had me do a proof of something. I don't remember what it was. But I did it on a big piece of board, you know? And he hung that on the wall. It was a nice thing to do. It made me feel good.

A third-generation mathematician described the impact of a demanding teacher in a school where the expectations were not generally high for students:

> Well, I could say definitely in junior high and high school, I had two Black female math teachers, which was really unusual, so I think that had a huge effect on me. One teacher that I had, I'm still really close to today. She was just really demanding—you know, I went to a public school where the math wasn't that good and the expectations weren't that high, but this one particular math teacher was really demanding and really exacting, expected a lot from us. People really did live up to her expectations. It just went to prove that people rise to the occasion, if you demand good things of them. I think she had a big effect.

Certainly for many mathematicians the mathematics experiences they had in school were memorable. All of the experiences were not positive—and indeed sometimes hampered mathematical development. As one second-generation mathematician who attended high school in the 1960s described his experience:

> In the 10th grade year, I took geometry and unfortunately I didn't have a person who knew geometry, so we pretty much taught ourselves. He was the teacher and we gave him due respect, but we had to do proofs. Geometry is not the easiest course. Of all of the courses, that is probably the most challenging one in high school. You have to prove this triangle congruent to that triangle given a set of axioms.
>
> The teacher didn't know. I would show him my proof and he would look at it and say, "That looks good to me." We would go with that. I don't know if they were right or not. So we made it through geometry.

Several mathematicians in this study participated in extracurricular summer programs that were designed for high school students as mathematics enrichment programs. Many of these arose out of increased national attention to and funding for mathematics, science, and technology programs, and several mathematicians directly reference this as resulting from the impact of the launch of the Sputnik satellite, the space program, and the Cold War.

These summer programs, often held on college campuses, gave mathematicians the opportunity to learn advanced mathematics and interact with other young people interested in math. One mathematician, who attended such a program at Yale in the 1960s, had wanted to take Algebra II in summer school so that he could take calculus his senior year. But the program interfered with his plans:

> I guess about three or four days into my instruction at Yale, I approached the math teacher. They had divided us up into four groups, and I was in the highest group as far as skill level. He was a secondary teacher at a public school in New Hampshire. I told him, you know, I really had expected to take Algebra II this summer, and I still wanted to do it, and I explained why. He said, "Well, I think I can teach you Algebra II while you're here." I would meet with him one-on-one two or three times a week. He said, "I'm just going to take you through the material that I taught in my Algebra II class." I was still doing all the other math, the math induction and the linear programming. I was really doing a kind of independent study. That was again a stroke of good luck that this teacher didn't say, "Well, you're learning all this sophisticated stuff and I don't think you have enough time to take Algebra II on the side." Instead, he took me aside and took the time with me. (Nathaniel Long, second generation)

Another second-generation mathematician attended a summer program as a high school student at an HBCU he eventually attended:

> This was about 1958, I think it was '58, '59, which was the second year that [the HBCU he eventually attended for his undergraduate education] had money from the National Science Foundation. I mean, everybody was giving money everywhere, throwing money everywhere for kids who were in math and science. Nobody cared about it before Sputnik. There was lots of money around to do stuff. That's how, I mean, Black kids from schools from North Carolina up to Philadelphia came to study math and science in the summertime. We all got paid money, I mean, $500 in those days was a whole lot of money, and $500 is what we were paid to do this. So the idea was that if kids got paid, they wouldn't have to work that summer. . . . We did some calculus, did series of equations, pretty advanced type stuff

> like that. Our book was by Dixon. Now we were high school kids so we didn't get very far in it, but the commentary was interesting. We did a little bit of algebra as well.

Finally, a third-generation mathematician who attended a science and math–themed high school recalls attending an engineering camp at the University of Maryland in the 1980s:

> My summer years were spent sometimes in sports camps, but between my junior and senior year, I actually went to a "minority engineering opportunity program," or MEOP. I went down to the University of Maryland for the summer. We took classes, and we were exposed to a variety of classes that a freshman engineering student would take. It was run by a Black teacher, and I can't remember his name. . . . He was actually from Baltimore, and he was one of the founders of the program. Those experiences [in high school] and those experiences during the MEOP program are the ones that are definitely the trigger points of my life. I decided that I was going to become an engineer. At the time, I knew nothing about becoming a mathematician. I was going to be an engineer. (Wade Trimble)

This is not to say that the most rigorous or enriching mathematics activities occurred during special summer programs. However, some of Black mathematicians' memorable experiences—"big steps" toward becoming mathematicians, as one said—occurred when they were able to study math beyond what other students in their school may have had the opportunity to do. Leroy Woods, who, like Wayne Leverett, Russell Means, and Eleanor Gladwell, attended a segregated Black secondary school in the South, remembers a teacher fondly:

> Probably the next big influential teacher might have been the teacher who came to my school to teach advanced math. And we didn't have advanced math at my school. So they pulled out maybe four or five students or something like that to make a class for him to teach. He was quite influential [for] me. . . . We covered material that I had never heard of. He was weird enough for me to want to emulate him and be weird like him—just known to be smart or thought of that way. He took me under his arm to a point where I decided I was going to drop the class. Because it was just going to be too much for

> me. And on my way to go drop it, I ran into him. He said, "So, where are you headed?" And I couldn't say, "I'm going to drop your class." So I ended up staying in it. Even the next year, and it was my senior year, there were only two or three of us in there. They let the class run. And by that time, I think I was ready to read the material before he lectured on it. I didn't quite know that was a step. I look back now and say, "Oh, that was a pretty big step." So there were a number of influential teachers along the way.

One second-generation mathematician, who was among the first group of Black students to desegregate his southern high school, recalls a teacher offering algebra classes in the morning:

> So what you would do is you would come in before school started and he would talk about algebra, you would work algebra problems and things like that. That, to me, was one of the formative experiences because that really made me feel confident when I got ready to go off to college.

Peer Contributions to Mathematics Success in Secondary School

It was not just teachers who were important to mathematics development. In many ways, mathematicians were able to meet other young people who were very interested in mathematics through opportunities provided by teachers. One mathematician who attended high school in the 1960s in a northern state describes having "very special" teachers, being on the math team, and studying advanced topics:

> They were very flexible, as I said, in terms of letting me study other things during the class. Then in the senior year I had two other friends, we kind of had a seminar that was loosely supervised by, I think, the head of the department. He kind of let us do our own thing. I mean if we needed we would talk about things from the book and do problems together. Stuff like that.
>
> Q: So these two friends that did this seminar with you, did you have other friends who were also interested in these kinds of things that were in the seminar?

> A: Well, first of all, there were very few minority students around. I mean I was one of the few in the school at that point. So these were mostly Whites, in fact, all of them. So it was also the group that was loosely around the math team. I mean we had a serious math team. I wasn't that good, but I competed. So one of the things about the math team was that you get together early in the morning and you discuss problems. And sometimes it's part of that whole kind of process, you just do problems.
>
> You know it's the whole time of Sputnik. I'm a beneficiary of Sputnik. . . . Of course I was born in [the late 1940s], but the end of the '50s is when you have all this attention to math and science. There were a lot of people there who were very much oriented towards science and so on, so it was very sort of nurturing. . . .
>
> Again, I was one of a fairly low number of Black kids in the school so since I was well prepared getting there and I was able to do the work, I kind of benefited from it. I had teachers who were interested in me and supporting me so I really can't complain.

Another mathematician who attended high school in the 1980s had a similar experience:

> I had who I think was a really good math teacher, and a really good group of students who sort of stayed in the same level math throughout our time in high school. And we just sort of, you know, it was a nerd club. We really sort of all—I don't know that we were thinking about pushing each other, but we were all sort of really excited about math, and doing well in math. So just by nature of that, we sort of pushed each other along. Got things organized so that, you know, it's not like we were working in groups or anything, but we sort of each—and it wasn't really competitive in terms of, you know, "Hey, I have to outdo that person," but it was always sort of we realized by that point that we were all pretty good at this stuff and that we could be pushed. And our teachers expected, you know, had really set the bar pretty high for us. And my parents always did too. So that combination of things, both having had the summer experiences, having a group of people, having a good teacher, having parents and the teachers set really high expectations, all those things really kind of created a nice environment for me to really know what I should expect of myself in terms of math.

One teacher did something rather unusual: she invited a mathematician's son, who was in 12th grade, to come speak to her eighth grade class and give advice about mathematics. For Stanley Parker, a third-generation mathematician who attended a Black high school in the South, this was an important moment:

> In eighth grade, our math teacher, Ms. Franklin, had one of the seniors, Dolan Falconer—Etta Falconer's[1] oldest son—come, he was in 12th grade. I remember she had him come to our algebra class one day. And I remember—he was just talking to us, again we were eighth graders, so he was giving us a little "old voice of wisdom." He said something that day, he said "if you're able to do everything, or if it's not going fast enough, work ahead, and do all the problems." And so I thought, "Oh yeah, I could do that." And it hadn't occurred to me and the teacher hadn't said to do this. . . .
>
> And she let me [do this], I didn't really appreciate it, but I do now. The time it takes to grade, the time it took for her to answer questions when I had questions. . . . I did every problem in that book, evens, odds. She checked my work. It made Algebra II sort of a drag, though. . . . Certainly it gave me a very solid foundation in algebra, but it was just one of those things where I had never thought of it, and someone [else] just planted the seed.

The previous vignettes describing school opportunities to learn advanced mathematics underscore the importance of peers in fostering mathematics socialization within school settings, across generations, regions, and eras. In chapter 3: Navigating the Mason-Dixon Divide, I explore more deeply Black mathematicians' experiences as adolescents in both predominantly White and predominantly Black settings and how they contributed to Black mathematicians' views of the world, their profession, and their responsibilities. More generally, it is important to note that peer interactions related to mathematics in school are not monolithically negative (with future mathematicians as social outcasts), nor are they monolithically positive (with future mathematicians being cheered as geniuses). Although there still seems a bit of a "stigma" around mathematics—"nobody wants to be a math geek," said one third-generation mathematician—the influences of school climate, peer norms, and teachers all contribute to how Black mathematicians in elementary and secondary school perceive themselves and are perceived by others. As Stanley Parker recalls:

> Q: How did your friends or other kids feel about your being tops in math?
>
> A: It's hard to say. It's hard to separate what was difficult from being good in math, good in school . . . it's hard to separate what was difficult and also being one year younger in my class (because I had skipped a grade). I was a nerdy little kid who [did] all the math. High school is hard.
>
> One thing I often remember, I was in eighth grade, I believe it was eighth—eighth or ninth—I was in high school, and I was on a little local television show, a little game show called "Intuition" but the "tu" was spelled numeral 2, so "In2ition." I was at the final round, if I had gotten my question right, I would have won.
>
> I remember getting to school the next day—and students knew to watch, I guess, they [school administrators and teachers] probably had announced it was going to be on—I remember, Vanira Peak, she was one of those who would just make your life miserable. She was loud, she was just one of those who would make life hard, just because she could. And I remember I was so stunned—Vanira Peak came up and said, "we were rooting for you! I said, he's in my homeroom!" and I was so stunned—she liked to take digs at everybody and me, especially, and she's up here rooting for me because I'm in her homeroom. And I remember that sticking out, she was proud of me, I guess, or something, and I could infer from other experiences that there was more of that than I realized.

Stanley Parker attended a predominantly Black secondary school in the early 1980s; Herb Carter attended a more racially diverse high school during the early days of its integration in the early 1970s. Carter recounts an interaction with a fellow Black student in his classes:

> He pulled me [aside] and he says, "Look, you have a responsibility." I still remember to this day he says, "You know, you have a responsibility. You're better than any of us in terms of doing this stuff." And he says, "You're probably better than most of the White students." Which at the time he said, "You're the number one student in this class. It's you, it's this bunch of White students, and then it's the Black students on the back end." He says, "You've got to stay number one, and you also have an obligation to help, you know, to tutor

> and stuff like that." So you know, anybody that was kind of interested I would help them tutor. Not because of him, I would have done that anyway. But I did feel this obligation because he would monitor what I was doing; he was a real character.
>
> Q: Was he older than you were? Same age?
>
> A: No. He was the same year, same age. He did a lot of leadership type of activity. You know he organized boycotts and he organized protests. He continued to do that after he graduated. He was really active in social issues. Like I said, he told me that that was my obligation, and I kind of believed him. There were times when I kind of didn't feel like studying, I would kind of like hear his voice. Which was really kind of interesting to me. And I wouldn't remember his name or his face if he would walk up to me now at all. In a way I would like to thank him.
>
> Q: What grade was this in?
>
> A: This would have been 11th grade.

Very few mathematicians described peer communities outside of school in adolescence that supported their mathematics development. Nathaniel Long, who grew up on an ethnically diverse block in his mother's Northern hometown, is an exception:

> One of the kids, Larry, was about five years older than me and he would play ball with us younger kids. I would play chess with him, and he started giving me these little puzzles. He was very interested in mathematics. He ended up majoring in mathematics and became a secondary school math teacher. I was the oldest kid in that group—this group of kids that were really second generation on that street. So I was the one that was older and perhaps more interested in mathematics than the other kids. (Nathaniel Long)

This peer community, however, existed within a broader neighborhood that Long describes as influential to his development as a mathematics student:

> I grew up in a—well, call it Little Italy—a mostly Italian and some Irish and German, but all Catholic [neighborhood]. And then there

> were a smattering of Black families. My mother actually grew up [on the block] the generation before, so they all knew each other. It was very close-knit. . . . It was a pretty bright group of kids. Their mothers had grown up with my mother on that street and they were all college educated. . . . They're still living there. But there was a sort of intellectual atmosphere—not explicit, but they all read a lot. They were very well-read and my mother talks about the boys next door, the twins, that stopped right in the middle of mopping the floor and they were reading books and that sort of thing.

Conclusion

The kinships and communities to which Black mathematicians belong in childhood and adolescence provide critical opportunities for their mathematics learning, exposing them to rigorous mathematics as well as supporting the development of fundamental skills. In addition, these kinships and communities serve to induct Black mathematicians as young people into networks of individuals and groups who are committed to their mathematics success. Rich descriptions of opportunities for mathematics socialization emerge from their narratives—and these occur within family conversations and activities as well as in mathematics classrooms and extracurricular clubs (formal and informal) within schools. Of particular interest are the family members with keen mathematics interests who for a variety of reasons were unable to fulfill their potential. There is a poignant refrain of "what might have been" in these narratives. These interactions with family members, teachers, and peers carry substantive weight in Black mathematicians' reckoning of key events in childhood and adolescence that contributed to their pursuit of mathematics later.

Most of the mathematicians interviewed have fond memories of their school communities in childhood and adolescence. For some, these communities resembled family networks, with warm, caring, and effective teachers who were actively engaged in ensuring that students' mathematics talent was developed and nurtured. This is particularly true for most of the mathematicians who attended segregated Black schools in the 1940s, 1950s, and early 1960s, although the picture is less rosy for some. For others, although the relationships may not have been as close-knit, opportunities to do mathematics were available, and mathematicians were encouraged by teachers to take advantage of them.

For mathematicians who were among the first, or the few, to desegregate all-White schooling environments, there is some ambivalence about those experiences, as we will see in chapter 4. Opportunities to engage in high-level mathematics were present for the mathematicians in this group, but a number of them question the availability of these opportunities for the other Black students who attended school with them. Although these mathematicians were supported in many cases by other Black students, there were key elements of isolation in their narratives. In these environments, peer communities did not always rise to the level of kinships, even for younger mathematicians. Within predominantly Black institutions, there appear to be stronger relationships with other students. But only 5 of the 20 third-generation mathematicians interviewed for this study report attending largely all-Black secondary schools.[2]

Although it is important to describe the structure and characteristics of mathematical relationships with peers as I have done here, the context of *where* and *when* these kinships and communities emerge is also critically important to explore. It becomes increasingly obvious that the region of the country as well as the time period in which mathematicians came of age has a dramatic impact on the kinds of interactions they have related to mathematics and with whom they have them. In addition, as I will demonstrate in later chapters, these kinships and communities have an impact on how mathematicians view themselves and their roles. In the remaining chapters, kinships, in the guise of strong, close peer communities and professional networks that provide structure for support and mentoring, largely outside of traditional mechanisms, continue to emerge as key to the lives of these Black mathematicians.

THREE

NAVIGATING THE MASON-DIXON DIVIDE

In this chapter, I use the notion of the Mason-Dixon Line—a now hypothetical but at one time very literal divide between North and South—as a metaphor for the border states in which Black mathematicians have frequently found themselves. These borders—physical, sociological, and psychological—have a significant impact on the mathematical lives of Black mathematicians. I begin with the literal geographical divide between South and North, largely by exploring the importance of the South as an incubator for mathematics talent among Black Americans, as previously described by some mathematicians who came of age there. The South has a particular connotation in American notions of liberty and opportunity as the site of rigid and violent racial repression, but it is also the site of strong Black educational and cultural institutions.

I then explore this divide figuratively, by discussing Black mathematicians' navigation between Black and White institutions, focusing on those who desegregated, or were among the first to desegregate, formerly all-White secondary schools and colleges and universities. Finally, I describe the internal, psychological boundaries that Black mathematicians are compelled to navigate beginning early in adolescence. In describing these psychological boundaries, I discuss how Black mathematicians' racial identities (Cross, 1991; Helms, 1990) are related to their mathematics identities. As Martin (2006) suggests, one's mathematical identity "encompasses the dispositions and deeply held beliefs that individuals develop about their ability to participate and perform effectively in mathematical contexts and to use mathematics to change the conditions of their lives. A mathematics identity encompasses a person's self understanding as well as how they are constructed by others in the context of doing mathematics" (p. 206). Martin and others (e.g., Boaler & Greeno, 2000; Nasir & Saxe, 2003) see mathematics identity as continually under construction and as being rooted in cultural, community, school, sociohistorical, and interpersonal contexts. Several mathematics education researchers (e.g., Berry, 2008; Hand, 2010; Martin, 2006; Nasir, 2000; Stinson, 2006) have demonstrated that mathematics identities, particularly for students of color, are formed, re-formed, and affected by dominant social, political, and media discourses concerning race, achievement, and mathematics. U.S. Black mathematicians, as members of a minority group within a small professional subset of individuals who are predominantly White and male, often find themselves navigating within largely White environments throughout their mathematical lives. For Black women, who are even more of a minority within the context of the professional community of mathematicians, these issues become magnified—but, as it turns out, later in their educational careers. They face a compelling divide in that their gender becomes increasingly salient to how they are perceived as mathematics doers.

Thus, this chapter focuses on the experiences Black mathematicians have during early and late adolescence—when they are in high school and college. As we will see, identity formation during these years is a profound and fundamental part of their later professional identities and how they navigate graduate school and professional environments, to be described in chapter 4: Representing the Race.

Coming of Age in the American South: "We Were Negroes Then"

More than half of the mathematicians interviewed for this book have roots in the South—a region that has been a site of both oppression and opportunity for Black Americans. As described in the preceding chapter, many Black mathematicians' formative experiences with mathematics—in one-room schoolhouse classrooms, in all-Black and predominantly White educational settings, in rural areas, towns, and cities—took place in the South. Despite state-sanctioned and rigidly enforced policies of discrimination and inequality for most of the United States' history, the majority of Black colleges—which 18 of the 35 mathematicians interviewed for this book attended—are located there.

Even for Black mathematicians who were not born or educated in the South, the South is often a key reference point in how they experience and make sense of their childhoods and their paths to becoming mathematicians. As David Blackwell noted, Centralia, Illinois, where he grew up was "*not* north of the Mason-Dixon Line"—not quite South, with its virulent racism and violence, but not quite the land of the free and the "promised land" of the North, either. Mathematicians who grew up in Maryland or Washington, DC, also describe those locations as "neither South nor North"—"even though Washington was considered South, you know, it wasn't southern like the real South," noted Evelyn Granville. The District of Columbia, in particular, offers an intriguing example of the curious contradictions of racism and opportunity. As the nation's capital, which in theory was to embody the promise of liberty and justice for all, Washington's long-standing policies of segregation meant that most DC residents were rigidly segregated by race in housing, schools, and jobs. Despite these practices, which were designed to underscore Blacks' status as second-class citizens, there were significant sites of Black excellence. As the home of Howard University, "the Black Harvard," Washington, DC, attracted Black faculty and students—strivers all. Many remained in DC after their time at Howard, whether as students or faculty, was completed. In addition, the sheer number of jobs in the federal government—even during the era of rigid segregation by race—was a lure for Blacks from the South and contributed to the development of a Black middle class. As Evelyn Granville recalled:

> Both [my mother and aunt] worked at the Bureau of Engraving and Printing in Washington, DC. That's where they make the money and stamps. And they both worked there, for years and years and years. And it was considered, you know, for African American women, of course we were Negroes then, you understand, that was a good job.
>
> In fact, there were two places that hired Negro women. One was the Bureau of Engraving and Printing, and the other government agency was the Government Printing Office. And those were the two main government employers of African American women. Men, of course, were messengers. And I did have one cousin, one of my mother's first cousins, who was a secretary in one of the government agencies, I don't remember which one. And of course that always, when I look back, that surprises me. How did she ever get that job? She was a secretary in one of the agencies, but that was very unusual. . . . So I always felt very fortunate that I was born and raised in Washington, DC, because it offered so many—what I think were extraordinary opportunities.

Gaining access to some opportunities but not all to which they were entitled as citizens, mathematicians who grew up in these border states recognized that they were unique to a certain degree and were beneficiaries of relatively relaxed social mores compared to their counterparts who grew up in the segregated South. Before the Civil Rights Movement reached its zenith, and certainly before the *Brown v. Board of Education* decision of 1954, for example, David Blackwell attended the local Centralia schools, which were largely White, in the 1920s and 1930s.

In the previous chapter, several mathematicians described their early educational experiences in all-Black segregated schools in the South. For many, the rich mathematical environments they experienced there were critical to their development and set them on the path to becoming mathematicians. As previously described, Wayne Leverett's, Eleanor Gladwell's, Leroy Woods's, and others' teachers gave them opportunities to learn additional advanced mathematics material, often within unrelenting systems of segregation that ensured their schools had fewer state-provided financial resources than the neighboring White schools.

But at least one other second-generation mathematician pointed out that doing mathematics in such an environment was not the paramount thing on his mind—the point was to excel and to get out of the bracing poverty his family experienced in a small town in Georgia.

> Well, the issue was not so much about my math experiences at that school. My family was what you would call poor, poverty stricken. And my brother and I, who were only about 19 months apart and we were one year apart in school, we basically saw education as our only way out of poverty. We weren't concentrating on math or science, et cetera. Our issue was to concentrate to do well to get out.

As Theresa Perry (2003) and others have noted, Black institutions, particularly in the South—churches, schools, and colleges—existed as a counterpoint to racist beliefs about Black intellectual inferiority. This is made very clear by Evelyn Granville's remembrances of the Dunbar School's teachers, highly educated themselves—one had a PhD in English from Yale, and several others had earned degrees from Ivy League institutions—who pushed her and her fellow students and encouraged them to go to college and then graduate school. Within these settings, these mathematicians were encouraged to be all they could be, and there was no expectation that they were lacking in anything—other than the opportunities granted to their White counterparts as a matter of course. These "valued segregated schools," as Vanessa Siddle Walker describes, were strongly supported by parents and community members and provide additional evidence to an already substantial knowledge base about the commitment of Black Americans to education in environments that sought to thwart this aim. When state and local political bodies saw no need to provide high schools for Black American citizens, Blacks founded and supported their own secondary schools. Worthy Black students went to great lengths to attend these institutions. As Clarence Stephens remembers,

> The last three years of my elementary work, I went to two country schools, little country schools. They didn't have any high schools I could go to; that's why we went to Harbison.

The Harbison Institute, an all-Black boarding school where Stephens and his siblings attended high school, was the only high school for Blacks in that part of South Carolina at the time. Founded in roughly 1882, it provided education for Blacks in South Carolina until roughly 1958. As Stephens recalls, at the time he attended, beginning when he was 13, there were roughly 300 students there, ranging in age up to 21 "who shouldn't have been there, but you know how it was—it was the South."

Although these mechanisms for educational opportunity for Black adolescents were significant, in numerous cases they were not permanent. The case of the closing of the Ware High School in Augusta, Georgia, in 1897—the first public high school for Blacks in the state, founded in 1880—provides a vivid example of the progress, stall, and regression that Black Americans seeking education experienced in the South (Anderson, 1988). Educational institutions for Blacks were constantly under attack by White power structures—from state legislatures to the United States Supreme Court—which questioned the need for publicly supported higher-level educational institutions (beginning with high school) for Blacks at all. As Rucker and Jubilee (2007) state, "the definition and goals of education became an important set of battlegrounds . . . for Black communities in Georgia and throughout the American South" (p. 151) and, arguably, nationally. Black educational institutions were often subject to fiscal challenges in terms of state and local support on the part of governmental agencies, as well as to interference and oversight by state and local officials with an interest in ensuring that education focused on manual training rather than the liberal arts or college preparation.

But it was not just excellent and highly valued elementary and secondary schools for Blacks that emerged in the segregated South. Historically Black colleges, with the exception of a few institutions like the first, Lincoln University in Pennsylvania; Cheyney State University, also in Pennsylvania; and Wilberforce and Central State University in Ohio, also arose there. As many have noted, these institutions existed within a paradoxical climate that refused to acknowledge the humanity or intellectual capacity of Black people. The Tuskegee-Hampton model for industrial and manual education was influential in determining what state officials and many philanthropists viewed as "appropriate" curricula for Black students, who, in their view, were destined to be blue-collar workers. Despite the prominence of this model and the rigidity of segregation practices in southern cities such as Atlanta, Nashville, and New Orleans, institutions such as Atlanta University, Fisk University, and Xavier and Dillard Universities thrived and contributed to the creation of a professional and intellectual class of Black leaders and scholars.

Within Black mathematicians' narratives, their descriptions of the lack of opportunity available to some of their forebears (parents, grandparents, other family members) in the South stand as a counterpoint to their own experiences. Black mathematicians of younger generations even today feel very strongly that the sacrifices and lack of opportunities experienced by some of their mentors and family members merit their paying back these debts and obligate them to help others to succeed in mathematics. For some, the fortune of having been born at a critical historical time, when opportunities were "opening up" for Blacks—even in the South—is clear. For others, the influence of teachers, professors, and administrators who recognized their talent and ensured access to postsecondary institutions they knew would be welcoming, or at least relatively welcoming, to young Black scholars was critical.

After Black mathematicians graduated from White institutions—including Evelyn Granville from Yale, David Blackwell from Illinois, and Clarence Stephens from Michigan—their employment opportunities before the 1960s were largely limited to historically Black institutions. Some mathematicians who had never truly lived in the South found themselves taking jobs at southern Black institutions, because these were the only postsecondary institutions that would hire them. Within these institutions, however, they inspired Black students to pursue additional education in mathematics. (Their experiences teaching in those institutions will be described in chapter 5).

In short, although opportunity in the South was in many ways constrained and circumscribed, this was the place where many Black mathematicians came of age educationally and professionally in all-Black settings. The experiences and opportunities there prepared some mathematicians for the later challenges they experienced in graduate programs and the profession.

"Planet Earth"—Navigating Borders in High School

There are some mathematicians, however, who had especially noteworthy experiences in that they grew up at a time in the South when they could benefit from the promise of *Brown v. Board of Education* and from the gains of the Civil Rights Movement. For some, this meant that unlike the previous generation of students, they had the chance to attend newly desegregated schools for their secondary education. Many mathematicians in this category first had elementary education experiences within close-knit all-Black communities, in vivid comparison to their later schooling experiences during early adolescence, when desegregation first occurred. Their experiences in these settings are compelling, as they have the perspective of students who shifted from being in all-Black settings to being in a new world: the desegregated high school.

In some ways, the promise of *Brown v. Board of Education* in exposing Black students to greater educational opportunity, and mathematics opportunities, in newly desegregated environments was realized. But Walt Austin's experience, described in the preceding chapter, in which he was among the first Black students to desegregate the local White secondary school and was invited to attend a morning Algebra enrichment program hosted by a teacher, was not shared by everyone. The costs of these opportunities could be high. Black mathematicians who desegregated White institutions noted that opportunity for most of their peers was tantalizingly out of reach, and being at the vanguard of desegregating these environments was not without some upheaval. As Stuart Potter, a second-generation mathematician, recalls:

> The community that I lived in was called Cherry Hill. . . . It was, again, an all–African American community. In fact, in growing up there, it was a separate world. I hardly even saw White people. Any time I saw them it was when they would come in the neighborhood to sell insurance or whatever. So going to Southern High School was quite a change. . . . I try to express that it is like you are here on Planet Earth and then every morning, you go to this other planet where you will be among these other people: White people. I had

no real social interaction with them because they were just barely tolerating us being there. In the class that I was in, I was in the so-called enrichment class, which meant that I was the only African American in that class. As I said, to me, it was so strange. It was like every morning going to school was like going to another planet and being with these other people. Of course, in [due course] you get to come back home.

Q: Why do you think you were the only African American in this enrichment class?

A: [sarcastically] It was an enrichment class. It was for people that thought. There, you know, Black people can't think. So I was really considered an oddity being in that class. Even in that state, you are kind of aware that you are competing with them and you are kind of aware of the fact that they are supposed to be superior and you are naturally inferior.

Potter's description of normalcy, that is, returning to his Black neighborhood as "Planet Earth," is shared in some ways by younger Black mathematicians who attended high schools in the 1980s and 1990s in terms of how they describe their negotiation of borders between their predominantly White high schools and their predominantly Black neighborhoods. This theme of going from being one of many and feeling "at home" to being the "only" and feeling isolated and out of place is a theme that frequently reemerges in Black mathematicians' narratives when they describe their schooling as well as their professional experiences. A third-generation mathematician who attended high school in the late 1970s describes his experiences in a predominantly White school:

Getting to high school I wasn't really in the upper classes, but I did take the science track, the college track, and I found that I did well in the college algebra. And what it was was that most of the Black kids in our school were all put in certain classes and then the few bright students were in the classes with the majority of the high school population. And I found I was in that situation in math class.

Potter's "Planet Earth" school experience was also the experience of a younger mathematician, Elizabeth Ricks, who earlier described her upbringing in a warm, close family and school environment in a predominantly Black southern neighborhood. When her family moved to a neighboring state for a few years, she was required to leave her predominantly Black high school for a predominantly White one, a "transition" she found "traumatic."

> I know that I did not work as hard in high school as I had before and not as hard as I did when I went to college. So I don't really remember too much of my math classes those last three years of high school. I don't think I did any more competitions. I don't remember too much about my last three years of high school, but I knew that I still loved math and I knew that I still wanted to do it. There was no question of what I was going to major in when I got to college.

These descriptions of shifting between nurturing and welcoming all-Black educational environments to all-White environments as alien, isolating, and traumatic are not uncommon in racial identity literature. Despite the promise of *Brown* to integrate the public sphere, the reality of continued discriminatory policies and practices in the United States regarding housing meant that many neighborhoods remained, and still remain, segregated even in the late 20th and early 21st century. Thus, many younger Black mathematicians who attended predominantly White institutions, while largely positive about their schooling experiences, also described themselves as going back and forth between their school worlds and home worlds. Their closest friends were in the neighborhood, not at school, and different mathematicians had varying degrees of success reconciling the two worlds. A third-generation mathematician's juxtaposition of "university high schools," which were deemed to prepare students for college, and her "neighborhood schools" is not uncommon:

> There were two what they call university high schools, which are schools that are geared towards people who are going to attend college. So I was bused in. None of the schools that I went to were my neighborhood schools. I didn't go to school with any of my friends in my neighborhood.
>
> Q: How was that? How did you stay connected to your friends in the neighborhood?
>
> A: I still hung out with them. I don't know. I am very much still me, so I still hung out with them on the weekends. Then I began making friends within my high school. Most of my friends now were in the

same program in high school. My neighborhood friends were older. We definitely aren't in contact as much as we would have been had I attended the neighborhood schools. (Laverne Richardson, third generation)

When asked if and how their friends, in and out of school, contributed to their math development, most younger mathematicians made distinctions between "school" friends and "neighborhood" friends. With the exception of a few mathematicians, most did not talk much with their close friends about math, and their fellow students in math class were not their close friends. As one mathematician recalled, he did not "have much contact with them [math class friends] outside of school."

Even in contemporary times, there are differential resources allocated to the schools that Black and White students attend. Schools that Black students attend in the 21st century have become more and more segregated, and the issues associated with racial segregation of schools—fewer resources, with less-qualified teachers who are less likely to be certified in the subject of mathematics (Flores, 2007; Walker, 2007, 2012a)—have an impact on the opportunities students have to learn rigorous mathematics. Nowhere is this made clearer than in this story, told by a third-generation mathematician who earned his PhD in the late 1990s, about his predominantly Black high school competing in quiz bowls against predominantly White, more affluent schools in his southern school district:

> Actually, it was kind of embarrassing, [with] regard to how prepared the other schools were, especially schools with some of the better resources. We would walk in as is. We'd go to practice, we'd practice some problems. We didn't have the best calculator out there at that time. There were calculators out at the time that we didn't even know were available that would help with just a two-dimensional graph, a polynomial equation. We pretty much just had your basic, solar-powered calculator. So that was kind of a humbling experience, what we did in the tournament as a team. Every once in a while, we got some individual problems here and there, but as a school, we didn't win anything.

As intimated by Potter and others, opportunities to learn within predominantly White schools are not always equitably distributed either. Black mathematicians reported that when they were attending predominantly White high schools, their parents had to be vigilant to ensure that their children were able to take advantage of high-level mathematics courses and opportunities. Two—the first from the second generation, the latter from the third—share their stories below:

> Now there were a lot of things that were going on that I didn't know at the time. For example, I think there was some objection to my getting into a special class in junior high from some White parents, which I guess she [his mother] was able to overcome and my sixth-grade teacher was able to overcome. So there was stuff going on in the background. But for me, I was just one of the few Blacks who was able to really tap into all of those resources.

> Well I have to say, again, it was really my parents who really took the lead on that. They never felt like they could rely on other people to push me appropriately. So, you know, I was enrolled in every gifted and talented program in the neighborhood or that the school did. And then at one point I went off to private school at I think seventh grade or something like that. And when I got there, I was enrolled in some low-level math and they thought to say, "What are you doing? Get him in the highest level math possible." So I sort of feel like they scrapped and fought [for] me to be in the highest level math wherever I was. And I did well at whatever they pushed me to be in.

Some younger mathematicians who attended predominantly White secondary schools shared vivid memories of teachers in advanced classes who expected them to be excellent students and held high expectations for their work within those settings. For example, one third-generation mathematician moved herself from pre-algebra to algebra mid-year when she felt that her pre-algebra class was not challenging enough for her.

> Then the algebra teacher was this Black guy, Mr. Bell. And so I was like, "Ooh, I want to be in Mr. Bell's class." He had this funny accent. He was very proper, and everybody had Mr. Bell jokes and I was just out of the loop. Yeah. I was like, "I need to move up to algebra." And I did horribly. I mean like he called me up to the board one day. I had no idea what was going on. You know, because I just got

> in the class and there was only like three Black kids in the class, so he was always, you know, kind of on us to be like, "Y'all need to be up on it." And I was always like, "Well, you know I just moved from pre-algebra."
>
> I had B's and stuff. But it was cool. His class was really fun. He was very tough.

She also recalled fondly another teacher, who taught her calculus.

> [He] was just so cool. He was an older White guy. And he would send us to the board to work math problems and we would get extra credit. [On one occasion] I was the last person at the board, and I was sitting there like sweating bullets. I was just like, "I can take the derivative. It's not difficult." And so he was standing there. He was like, "Oh, Francine. That looks good from this angle. We just need to get one of those Algebra II students in here to work out those little details. Don't worry about that. You're doing calculus now." And I was like, "Yeah, that's right. I'm doing calculus. Yeah, this is hard stuff." He would have us feeling so pumped that we were in the AP calculus course.

Several mathematicians describe their mathematics teachers as being particularly charismatic people who made math interesting and fun:

> But by far, who I am as a professor now and who influenced me the most was my pre-calculus and calculus professor. His name is Stan. We had to call him Stan. We could not call him Mr. or he would literally deduct points. He was very animated. It was the first time that I saw somebody who really enjoyed math. Up to that point, I just liked math and was like, "Okay. We will do it." But this person would get excited about trigonometry and sine curves. You had to make them smooth and flowing or he would deduct points. So he is by far the most influential person in my life. In fact, I am not sure I would be a mathematician if I didn't run into him in the 11th and 12th grade.

Interestingly, these two teachers seemingly have differing philosophies about teaching math—one is very concerned with the mechanics and format of students' work, deducting points for seemingly minor issues, and the other is able to recognize the quality of his student's calculus work even if the algebra details were not perfectly worked out.

Teachers of other subjects also had an impact on Black mathematicians—as a second-generation mathematician relates, her chemistry teacher was very influential in her junior high and high school careers:

> And even though this man was not a mathematician, he was a chemist, he very, very strongly influenced my career as a mathematician because he was actually, in retrospect, I think he's probably very good at mathematics. He had gotten a master's degree in chemistry. . . . And so I had him as science club advisor in seventh and eighth grade and then he was a science teacher in the eighth and ninth grade and then homeroom teacher for me. And he is the one that actually pointed me in the direction of going to [a specialized high school in science], which I didn't even know existed. . . . He encouraged me to, and in fact I think played a strong role in getting me enrolled in something called the Saturday Science Honors Program associated with Columbia. . . . And what it did is every Saturday it had various courses, enrichment courses for students in New York City. And it had a program in the summer as well.

Many mathematicians credit these experiences—whether they attended predominantly White or predominantly Black schools—with charismatic and caring teachers, in extracurricular and enrichment opportunities with peers, with their becoming mathematicians. Building on early school experiences and key mathematics experiences with significant family members, Black mathematicians' secondary schooling experiences with math continued an important socialization process, even before they entered college. Although most mathematicians report very positive experiences with mathematics and mathematics teachers, several point out the limited nature of the secondary school mathematics curriculum within mathematics classes, the difficulty in ensuring proper course placement in advanced mathematics classes, and feelings of isolation in some mathematics settings. Black mathematicians felt "set apart" in multiple ways from their classmates—for most it seemed uncommon to do well in mathematics, regardless of whether they were in a predominantly White or predominantly Black high school, and when they were in advanced math classes in White schools, they were often the only Black student, or one of a few Black students in the class.

"Don't Join the Band": Mathematics Experiences in College

Many high school experiences—feelings of isolation, the importance of strong teachers and/or faculty who had high expectations for students, exposure to extracurricular mathematics activities—were still very salient for mathematicians, older and younger, once they entered college. Only a few mathematicians interviewed for this study did *not* major in mathematics in college. Most of those who did major in mathematics had relatively easy paths through the major, although a few were discouraged from pursuing it or actively encouraged to change their majors. In college, most had the opportunity to meet other students who were as interested in math as they were and to take classes with and work with faculty who were committed to their joining the profession.

"I Just Loved It All The Way Through": Becoming a Math Major

For many mathematicians, their early very positive experiences with math at home and in elementary and secondary school ensured that there was no question about what they would major in in college. When asked how he decided to be a math major, Stanley Parker recalled:

> Sure, that was from the beginning. I just always loved math, even from the [earliest] time and so I always just went forward with it.

A second-generation mathematician who attended segregated schools in the South noted that unlike other subjects, he liked that math was "absolute":

> [In elementary and secondary school] I would tinker with things like model airplanes or telescopes or scientific gadgets. And they amazed me because there was order to how they were put together and how they functioned. When I got to college it was clear to me that I had somewhat of an analytical mind. And so I wanted to do something in math or science. And that I did not want to go into English or social studies or history because every time I picked up a different book everybody wanted to change the story or the rules. They didn't change in math or science.

Others, as noted earlier in this chapter, were strongly influenced by teachers in school who showed them the beauty of mathematics. Leverett, one of whose teachers had allowed him to teach the class, reported that he hadn't thought about being anything else than a math major: "I wanted to be a math guy. Every time I saw my high school teachers teaching it, I knew that one day I wanted to be up there explaining how to do these algebra problems." Most of these, as strong students performing well in high school, were strongly encouraged by their teachers to major in math and to attend certain colleges. According to one of the Morgan State mathematicians, his teacher had attended some sort of summer science institute for high school teachers at Morgan State.

> He was so impressed with the math department [at Morgan]. When [the teacher] found out that I was going to Baltimore for the summer [to visit my father] he said, "I want you to go out to Morgan State and tell Dr. Stephens that I said to give you a scholarship in math."

Several mathematicians, including David Blackwell, Wayne Leverett, and Laverne Richardson, had assumed they wanted to be teachers, because that was all they knew that could be done with mathematics at that time. David Blackwell's career plans when he entered college were to become an elementary teacher, as he had been promised a job by a school official when he graduated from college. In the 1930s and 1940s, that was not an insignificant consideration. Laverne Richardson, who benefited from a charismatic and excellent teacher, Stan, during her senior year of high school in the 1990s, first entered college as an engineering major but soon changed her mind:

> I did realize that I wanted to major in math, but I changed my major from engineering to math education. Why? Because I thought, "There is no other option. You can't just be a math major. You must have to be a high school teacher." I knew that I liked math and I thought there was no other option but to teach high school.

This notion of not knowing the career opportunities associated with mathematics was a theme throughout several interviews.

During the interviews, mathematicians who had not initially majored in math reported switching majors (from engineering to math, psychology to math, music to math, etc.) for a variety of reasons.

> I started off thinking I was going to a pre-engineering program, because it's a liberal arts school for engineering majors. But I was

having a great time as a football player and in college life. Then I decided to switch over to computer science, which was an available major. I remember distinctly, my sophomore year, that every time I would hit "run," the computer never ran. I realized, "This isn't for me." I went to the career center, had a discussion with one of the counselors, and just didn't realize the career options in math. She was able to show all of the things that were possible with a math major, or that, you know, fields that recruited math majors. That was an eye-opener, really my first look. That's when I switched over to math. (Claude Franklin, third generation)

A couple of years into college I realized I didn't like biology all that much. I didn't think very much of math in terms of a major just because I didn't really know what there was actually, beyond, you know, calculus and things. I actually was pretty naïve about that—I didn't know what a mathematician was or what they did. But it was really the thing I knew that I could finish and that I liked. And I became interested probably when I took [abstract] algebra. I liked the idea of what I had to prove there and those kinds of things. (Richard Reynolds, third generation)

Two mathematicians were initially music majors for whom mathematics courses were initially "electives."

My major was music, piano. And I was taking mathematics because I liked it. And also because I'm not sure if I was going to stay in music. . . . As I went through my [college] courses I would have conversations about math and music with my teachers. Eventually one told me that I was not going to be a concert pianist so I would probably make a better living in mathematics as opposed to music.

I started as a music major, and I was doing something that was kind of odd; I was just joking with somebody about this. I took math classes because they were easy A's. You know, you got to pack your schedule, your electives with things that you can easily get A's in. So I padded mine with calculus. And then at some point, I said, "Logically speaking, music may not pay the bills. If I take more math classes, I can always have, like, this day job." . . . I took all these courses I was interested in. [Eventually] I was in striking distance of getting a

> degree in history, philosophy, and math, and the logical choice was math. I never got the music thing together, which is funny.

Another mathematician who eventually attended Morgan State, Scott Williams, recalled being very good in music as well as math:

> I had a scholarship to go to Peabody Conservatory, a fairly famous musical conservatory in Baltimore. I had a scholarship there for saxophone, and a scholarship to go to Morgan for math. The math professor there was this famous guy named Clarence Stephens and he said, "Don't join the band." So I just put down all my instruments, that was it, I was done. I knew what my road was. I knew that PC was not the end of my life. I knew this before I even got to high school that I wanted to do research in math. I didn't know what it meant, but I knew that's what I wanted to do. So all the other things along the way were just stepping stones to getting where I could do research.

Engineering was mentioned quite frequently in interviews, because when mathematicians were told they were good in math by a teacher or significant adult that person usually also told them that engineering was the best career option. At the time that many third-generation mathematicians were entering college, there were significant scholarships for engineering majors but not for mathematics majors. This was the case for Laverne Richardson, Stanley Parker, and others.

> Everywhere I applied, I checked "math major," you know, I wanted to be a math major. And so I could get accepted anywhere, I just couldn't get any money. All the scholarships were in engineering. And I had some familiarity with engineering, I'd gone to a one-week summer program [in high school], and you know, people saying "Oh, you're good at math and science, you should be an engineer." And even now, to this day, they still say that. . . . And it's funny, you can be an engineer, but you can also be a mathematician. And I was fortunate at least that I knew I didn't have to be an engineer. I could, but I could still keep going in math. And I wanted to do that. So I just loved it all the way through. And then when I accepted the scholarship to do the dual-degree program [in engineering and math]

> it was because, yeah, I could get the scholarship and I could do that engineering stuff, but at least I could still be a math major the way the program was structured.

Other students, like William Burris, seeking to be engineering majors in college, discovered that their chosen institutions did not have the major, and so they chose the next closest thing: mathematics. Burris, whose father was a mechanic, reflects:

> I really wanted to be an engineer, but it's not that I really knew what an engineer did. It's kind of like, I had chemistry and physics in high school, and now I'm asking, "what profession pays the most money? Oh, chemical engineers get paid the most. Okay, I want to be a chemical engineer." But when I got to [college] they didn't have any engineering program at all. So I figured, what's closest? Engineering is a little bit of math, a little bit of physics. So I majored in math and minored in physics.

"The World I Lived In": Race, Place, and Mathematics

While the decision to major in mathematics for most mathematicians was a simple one, being in college and pursuing mathematics were, for some, inseparable from their racial identity. Enrolling in both predominantly White institutions and historically Black colleges and universities, Black mathematicians entered college and found that issues of race, whether in the South or North, were to be a part of their college experience.

Roughly half of the mathematicians interviewed for this book attended predominantly White undergraduate institutions. For first-generation mathematicians, who often attended colleges in the North, these institutions had already been nominally desegregated. However, as David Blackwell attests, there was still segregation practiced in these "border" institutions. As a Black student at the University of Illinois, he was not permitted to live in campus housing and, instead, lived in a building that housed Black fraternity students. Blackwell benefited from a fraternity brother's networks and know-how:

> You had to take a foreign language, and I had studied Spanish in high school, and I planned to continue that. But one of my fraternity

> brothers told me . . . that I should take German instead of Spanish. Because, he said, "If you're pretty good at mathematics, you may want to go on for a PhD. And Spanish won't help you, but you'll be required to read German, so you should take German." That was extraordinarily good advice from just one of my fraternity brothers who was two or three years ahead of me and could look that far ahead. (Wilmot, 2003, p. 12)

Other mathematicians of the first generation report that they had no problems related to race at their undergraduate institutions. Evelyn Granville, describing herself as "probably the poorest child that ever went to Smith College," said that she "never felt out of place. And maybe I was oblivious to things, but I never felt out of place at Smith. It never bothered me to be the only Negro in my class."

For Raymond Johnson, who was among the first Black students to attend the University of Texas, the situation was a bit different:

> So it was recently desegregated, but not quite because the dorms were still segregated. The athletic program for example was still segregated. There were no Blacks playing football, basketball, or any sport activities at all, any of the college teams. You could play intramurals, but not on the college teams.

Despite these practices, Johnson benefited from the fact that one of his high school teachers had gone to the University of Texas and helped to smooth the path for him there:

> On the other hand, [my teacher] got his master's from the University of Texas, and introduced me to his professor there. He realized that I was pretty good in math so he actually made me take reading courses instead of going into the regular calculus sequence and things like that. [He] had taught me a little bit of calculus and Dr. Curtis taught me the rest of calculus more or less on a one-on-one basis.

Leroy Woods, a second-generation mathematician, was the first student from his segregated high school to attend the all-White flagship university of his southern state. Excelling in his mathematics classes and wanting to change his major from engineering to mathematics, he was kept from being a math major there, as he describes it:

> I do remember going over to the math department to tell them I was going to change my major. I expected them to kind of jump up and shout, "Wow, we got another math major." And I went in and asked the secretary if I could see the chair and tell him I wanted to change my major to math and he could tell me what to do. And she went back to his office and she came back and said, "Well, what do you want?" And I said, "I want to talk with him about becoming a math major." She went back and talked with him again. And kind of came out. So, the conversation went with her relaying messages. So, she finally came out and said, "Okay. You can't be a math major." And I probably came out with a few words. "Well, I can't be a math major, it's going to be okay." But I never saw him. And so, as a result, I couldn't be a math major.

For some mathematicians who attended college just after the Civil Rights Era, whether they went to predominantly Black colleges or White colleges in the South, their mathematics experiences were inextricable from the incredible social changes that were occurring around them. Some of them had first-row seats at these movements, and they were active participants in protests themselves. Leroy Woods himself notes the pull of protests:

> At the time, it was late '60s, early '70s, I was only in college for two years. And that was at the time civil rights protests were going on in America. Everybody was protesting, beyond civil rights, the Vietnam War. America was in a protest state. We were doing the same here. And so, the protesting I did eventually led to my being expelled from [my undergraduate institution]. And I had maybe just finished my calculus sequence. And I think I was in differential equations when the final order came for me to leave. I was just about to get to the upper level of mathematics.

Some mathematicians talk about their perception that there was a conflict between doing mathematics and participating in some of the political movements that were going on at the time they were pursuing mathematics, either as mathematics majors in undergraduate institutions or as graduate students in mathematics at predominantly White universities. Some were very active, but others did not actively participate in these social movements. Two, Ryan Kennedy and Harold Prince, describe these experiences below:

> But [that was] the reality of the world I lived in, you know, from when I was young and living in segregated North Carolina with my

parents then to various battles with civil rights during my youth. I mean, I get a call from my father trying to buy Christmas presents for the family and he couldn't go into any stores. There were all kinds of things like that as a matter of fact. You couldn't try on any clothes in any store. What you bought was it. If you were Black, what you bought was it, no taking it back, no trying stuff on to see if it fits. If you want to try it on, you buy it first. Weird things like that. Not being able to eat. You could eat at restaurants in a Black neighborhood, but that's about it. There was no other place.

The students at [his HBCU] who wanted to go to the movies had to take a bus and two transfers to get to the movie theater if they didn't have a car. It would be over an hour ride just to get to a movie theater. So even though there was a movie theater that was only ten minutes' walk from [his HBCU], they couldn't go to that and the restaurants.

There was a lot of work for civil rights to do. When I was a student in college we protested that movie theater and what we did is that we would form long lines to buy tickets. You got there to buy a ticket, they said, "No. We can't sell you a ticket." But we get up to the line and a ticket cost, I don't know, if it was $3, everybody had $3 worth of pennies. So you'd count it out, even when they were told they couldn't get into the theater. They counted $3 worth of pennies out. So White people wouldn't even try to deal with it. They'd see the line there, and they just went away. We basically shut down that movie theater for like weeks, at least two weeks. That was newsworthy, it made *Newsweek*.

There were all kinds of things like that that were going on. I can't remember if I was a junior or senior when the first March on Washington happened. Political stuff like that certainly interfered with any math. You kind of were forced into doing something or other. . . . I mean, if you saw what was going on and causing you strife, you did stuff. That's how I was. . . . I was always doing that. I was in graduate school forming organizations when I was in grad school. Then as a mathematician, I was doing the same thing. (Ryan Kennedy)

So I went to college and other things happened. You know in the '60s I was very much kind of an activist and that kind of in some ways sidetracked me because I didn't go to graduate school immediately.

> That was not in mathematics. I kind of lost interest in math. Well, I couldn't really fully, you know, reconcile it [his political organizing and the mathematics]. There were things more important than mathematics; I think that's the second thing that I realized then and I continue to recognize it. Mathematics is great, but I think there are things that are more important than mathematics. So there is this tension between doing something that you like and doing something that's important. . . . The problems of the world are not mathematical, or even scientific. I think they are social and political. That's the hard thing, to convince people to make changes social and political. It's nice if they know mathematics and that is what I try to do in terms of my profession. You know I want people to learn mathematics. I hope that it will be useful, but I am not a person who believes that just technical knowledge itself will solve society's problems, I don't believe that. (Harold Prince)

Related to Harold Prince's point, there are modern philosophies and pedagogies of mathematics education that challenge the traditional stance of mathematics as a politically neutral enterprise. For example, Ladson-Billings (1997), Gutstein (2006), Leonard (2008), Tate (1995), and Martin (2009) describe mathematics as a tool for social justice and liberation to address societal inequities and conduct research with and about teachers who adhere to this philosophy. Further, Bob Moses's Algebra Project and its offshoot, the Young People's Project, seek to develop young people as advocates and agents for their own education, training them as "Math Literacy Workers," analogous to Moses's work as a civil rights worker in the 1960s (Moses & Cobb, 2001). All of these initiatives seek to develop young people's mathematical skills so that they can become more active citizens and advocates for their own learning by challenging existing and inequitable practices in schools and communities. These relatively recent developments in mathematics education are in some ways a response to lingering patterns of discrimination but do not reflect the kind of overt racism and discrimination that previous generations of mathematicians experienced in their everyday lives and that drastically circumscribed their opportunities. However, contemporary mathematicians report, troublingly, instances of racism and inequitable treatment in, especially, their academic and professional lives.

Mathematicians who attended White colleges in the 1970s or later rarely report overt acts of racism on those campuses but do describe instances in which they were challenged about being mathematics majors. This was mentioned most frequently by women. One third-generation mathematician recounts that

> There was one really well-renowned professor at [my undergraduate institution] who basically told me, "Pick another major," because I was a math major when I first started out. I wanted to be a math major. I later found out from a White female postdoc [that] he was definitely sexist. She didn't even go into whether he was racist or not, but it definitely came out later that he was definitely sexist and he still thought math should be a male thing. That was the main reason I didn't major in math in college. I was 17 and impressionable, and here's this big-name person telling me this, so I kind of believed it at the time.

Often Black students within these predominantly White environments, such as David Blackwell and his fraternity brothers decades before, created their own communities however they could. Some mathematicians describe instances similar to those they had in high school, where they had different, and discrete, categories of friends:

> You know, the truth is I think our department didn't do a very good job of inviting broad participation in those things. Combined with the fact that I think I was not particularly looking, you know, I don't even know that I knew to look for things outside even if we were offering them in the department, or they were offering them in the department at the time. And there again, it was another situation where I knew some people in those courses but there, even to a much, much greater extent than high school, it was like I was going there to do math, but none of the people who were there were people who I hung out with, or did anything with. There was a real disconnect from that world from who my friends were.
>
> Q: Why was that?
>
> A: Probably because one of the reasons I went to [this institution] is that I felt like it had a really terrific Black community. And I would say that was the real central core of who my outside-of-academic group was, who my hanging group was. That and I wrestled in college, and those also, my teammates on the wrestling team were also

sort of the other central core of the folks who I hung out with. So of those two cores there weren't, I don't think there were any mathematicians in that bunch. So the social and the academic didn't really, at least on that dimension, didn't really mix all that well.

In retrospect, I sort of recognize how much you miss if you don't have any mixing at all in that area. But something I think our department now spends more time trying to create environments where people can sort of be and develop community, which I think is important. But it either wasn't there when I was there or I was out of the loop. (Richard Reynolds)

Conclusion

Throughout high school and college, Black mathematicians experienced multiple opportunities to develop strong mathematics identities, in segregated schools as well as in desegregated environments. The variety of settings and situations in which mathematicians were doing mathematics in high school and college illustrates the importance of multiple access points to mathematics, both within and outside of school. For these mathematicians, particularly those who were among the first to desegregate their institutions, mathematics was inseparable from the social contexts of the time. In a few instances, challenges to Black mathematicians' mathematics excellence—through incorrect course placement or encouragement to change majors—threatened to derail mathematics talent, promise, and potential.

For some, there were influential adults (including teachers in secondary school as well as college professors) who helped them to consider becoming mathematicians. These experiences were foundational and formative. However, it should be noted that for many mathematicians, these influential adults were not all mathematics teachers or professors. Other individuals, including administrators, contributed to their development.

As these narratives show, there are times when Black mathematicians' racial identities and mathematics identities are inseparable, whether it is older mathematicians recognizing that certain opportunities may not have been available to them because they were Black or thinking about how their experiences desegregating White environments affected their development as students as well as mathematicians or younger mathematicians reflecting on their school experiences and the racialized experiences they had in mathematics class. These narratives show that as Black mathematicians come of age they become increasingly aware of how their racial and mathematics identities shape and are shaped by each other (O'Brien, Martinez-Pons, & Kopala, 1999; Martin, 2000; Mehan, Hubbard, & Villanueva, 1994; Moore, 2006). Narratives that Black mathematicians share about gaining access to rich mathematics in school, the importance of teachers who recognize their interest and/or talent in mathematics (sometimes late to develop, other times immediately apparent), and the experiences that Black mathematicians have within predominantly White environments, which cover a range of characteristics (including overtly hostile and racist, tolerant and tolerable, and nurturing and demanding, at times within the same institution), are obviously critical parts of their mathematics identity formation and influence how they view their graduate school experiences. (As will be described more fully in chapter 5, second- and third-generation Black mathematicians who attend HBCUs describe very different college experiences than those told in this chapter of predominantly White institutions.)

For many, there are certain critical transition points in mathematics—from elementary to secondary school, from secondary school to college, and from college to graduate school. These transition points mark stages when mathematicians could go "off track" and, indeed, mathematicians share stories of colleagues and friends—"equally bright or brighter"—who did not persist in mathematics at these critical transition points. The mathematicians in this book navigated these borders by drawing on family support and encouragement, established kinships and communities, teachers and other school adults, and their own strong sense of self. As the next chapter will show, however, when Black mathematicians enter graduate school, issues of race become more salient, and for Black women, the way becomes particularly perilous. As the quotation below makes clear, there is a marked shift for some Black mathematicians when they enter mathematics classrooms in postsecondary institutions:

> Q: Did you feel any other times that you were isolated from others in mathematics?
>
> A: No, I don't think I ever did. I was really lucky. I don't think I ever felt isolated, because I had a group of friends—as I said, [we] were in the AP classes [in high school] together, but even when we were in elementary school, we were all kind of . . . I don't want to say we were nerdy, but we all had an interest in math and science and all worked really hard. In college and grad school, I felt isolated, not because I'm not interested in math, but because I'm the only African American in the room, and sometimes the only female.

FOUR

"REPRESENTING THE RACE"

> In math class, if the teacher calls on you, and you're the only Black person in there, you don't want to say something that's stupid because she might think you're representing your whole race when you say it.
>
> —A 10th grade high school student (Walker, 1994)

The notion of "representing the race" figures prominently in the educational and professional lives of many Black mathematicians. For some, "representing the race" comes as a function of their being the only Black student in their high school mathematics classroom, as the high school student referenced above notes and as mathematicians themselves recalled in the previous chapter. Black mathematicians also may be at times the only Black person in college or graduate school math classrooms. "Representing the race" has also meant, for some of those interviewed, that they were among the very first Blacks to desegregate previously all-White environments, whether those were schools, universities, or professional environments. Their own notions of Blackness and any responsibilities they feel about being among the few Blacks in the mathematics field contribute to how they perceive their identities as mathematicians, researchers, university faculty, government employees, and industry professionals. These issues may also affect how they see themselves and how they are seen by others, as well as how they interact with others in the field.

Yet mathematics as a discipline, arguably more so than others, has a reputation of being "color-blind." If one demonstrates talent, the argument goes, the field will be open and receptive to that person. But from the 18th-century era of Thomas Fuller and Benjamin Banneker, Black mathematics doers have faced challenges about the depth of their intellect as well as limited opportunities to demonstrate their merit. Their achievements are usually understood within the context of their Blackness and, indeed, are often inseparable from it. Often seen as high achieving in mathematics *in spite of* the supposed limitations of being Black, Black mathematicians in the 21st century still have to navigate troubling racialized experiences in the field.

Brothers in Science

What we know of Thomas Fuller largely comes from an obituary published in 1790 in the *Columbian Centinel*, a Massachusetts newspaper, and from an article on "arithmetical prodigies" published by E. W. Scripture in the *American Journal of Psychology* in 1891. In that article, Scripture recounts Fuller's interactions with "two [White] gentlemen" who sought to satisfy their curiosity about the "Negro Calculator." They asked him to answer various questions: how many seconds there were in a year and a half and how many seconds a man has lived who is 70 years, 17 days, and 12 hours old. When Fuller answered the latter question,

> One of the gentlemen who employed himself with his pen in making these calculations told him he was wrong, and that the sum was not so great as he had said. (Scripture, 1891, p. 3)

Fuller replied that "the master" had forgotten the leap year. Fuller's *Centinel* obituary makes his status clear:

> DIED—Negro Tom, the famous African Calculator, aged 80 years. He was the property of Mrs. Elizabeth Cox of Alexandria. . . . This man was a prodigy. Though he could never read or write, he had perfectly acquired the art of enumeration. . . . Had his opportunity been equal to those of thousands of his fellow-men . . . even a NEWTON himself, need [not] have shamed to acknowledge him a Brother in Science.

Fuller's legacy as "property," not a citizen, serves as an example of what might have been. The notion of equal opportunity did not in any way apply to him because of his enslavement as well as his Blackness, as the obituary writers well knew.

Benjamin Banneker's "colour" always figured centrally in his successes as an inventor, craftsman, and mathematician. This juxtaposition of novelty (that a Black person could do such things), opportunity (both that Banneker was able to demonstrate his potential and that Banneker's work came to the attention of many Americans, including Thomas Jefferson and other prominent individuals), and intellect are present in others' descriptions of Banneker's work and continues to be present in discussions about Black mathematicians today. An excerpt from the preface of Banneker's 1796 *Almanac* highlights this juxtaposition:

> To whom do you think are we indebted for this part of our entertainment? Why, to a Black Man—Strange! Is a Black capable of composing an Almanac? Indeed, it is no less strange than true: and a clever, wise, long-headed Black he is: it would be telling some whites if they had made as much use of their great school learning, as this sage philosopher has made of the little teaching he had got. The labours of the justly celebrated Bannaker will likewise furnish you with a very important lesson, courteous reader, which you will not find in any other Almanac, namely that the Maker of the Universe is no respecter of colours; that the colour of the skin is no ways connected with strength of mind or intellectual powers. (Bedini, 1999, p. 339)

Although the writers of the preface suggest that the almanac is evidence that intellect has nothing to do with skin color, they point out Banneker's singularity. They also document that there is undoubtedly societal surprise that a Black person could complete such a work. Further, echoing Thomas Fuller's obituary, they acknowledge the differential opportunity in learning that is available to some Whites and Banneker, and how Banneker has made great use of what he has been "given."

More than 200 years after Banneker's and Fuller's times, these themes of race, novelty, opportunity, and intellect continue to figure prominently in both discourse *about* Black mathematicians and in Black mathematicians' *own* discourse. In short, for some Black mathematicians, their work—and their very presence in the field—is subject to discourses about race that may or may not be salient for others. The discourse can be self-imposed or imposed by others, and it can frame how Black mathematicians view their own opportunities and experiences. The discourses about race and racial identity that Black mathematicians may use in discussing their work are generated, I argue, by their life experiences and, in particular, their experiences in schools and in their communities. For example, mathematicians who attended segregated schools in the era predating the Supreme Court's 1954 *Brown* decision have a complex and nuanced picture of how segregated schooling helped and/or hindered their mathematics development. Further, given the extensive resegregation of U.S. schools in the 1990s, these experiences in largely all-Black elementary and secondary schools are not limited to mathematicians who earned their PhDs before 1965. Mathematicians earning their PhDs in later eras may have attended predominantly White educational institutions from elementary through graduate schools or may have attended predominantly Black schools for most of their school careers. They may have attended historically Black colleges and universities for their undergraduate studies and predominantly White institutions for graduate school.

As this chapter will demonstrate, issues of race and identity continue to have spatial and temporal qualities that influence mathematicians' experiences as they are inducted into the profession through graduate school and professional experiences. More than 150 years after Banneker's and Fuller's times, one young Black mathematician still had to grapple with the meanings of Blackness and citizenship.

"For the White Citizens of Texas"

For Raymond Johnson,[1] being Black meant that his options for graduate school in his home state of Texas depended on his being Black at the right time. Previous to his admission, Rice University had been challenged in court on the basis of its charter, which held that it had been established in the will of William Marsh Rice "for the White citizens of Texas." Johnson enrolled in 1963 as a "research associate," occupying a tenuous and nebulous position not unlike that of W. E. B. DuBois at Harvard half a century before—with the sensation of being *in* the university but certainly not *of* it (Levering Lewis, 1993):

> They had loosened the restriction on the citizens of Texas part many years before to make it accessible to people outside of Texas. But then when they decided to let in Black students they were trying to break the part of the will that referred to "White." So these two alumni sued to try to get the will upheld. So with that I was not a student the first year. I was instead hired as a research associate, a paid one. It was the same thing I would have been paid as a student. As a research associate I could go to class. I went ahead and took

> classes and just stayed there. The case was dismissed by the next year. So 1964 was the first year I actually enrolled in Rice. (Raymond Johnson, interview)

Johnson's enrollment at Rice meant the redefinition of what it meant to be a citizen of Texas and forced the state to confront directly unfairness, inequity, and injustice. Nowhere is the dilemma of being Black in a White world more telling than here—when the reality of second-class citizenship butts up against the idealized American dream of opportunity and equality.

Like Raymond Johnson, other Black mathematicians were among the first to desegregate White graduate institutions in the South in the 1960s and 1970s. Some of these included Robert and Sylvia Bozeman, two mathematicians who met as mathematics majors in college in Alabama and subsequently married, and Stuart Potter, whose descriptions of his high school experiences in a newly desegregated high school may be read in chapter 3. These three mathematicians, who had all attended HBCUs, noted that their going to graduate school held a certain significance for their undergraduate faculty mentors in terms of the mission of increasing representation of Blacks in the sciences.

Robert and Sylvia Bozeman had each other and at least two other Black students in the graduate mathematics program when they entered Vanderbilt University in 1968. Sylvia Bozeman described the transition from their undergraduate alma mater to Vanderbilt: "I finished college in 1968 and at that point I had never gone to school with a White person. I'm about to finish college, and I've never been in class with a White student. So I really came from an era of segregation." Robert Bozeman continued:

> The time period was 1964 to 1968 and Vanderbilt was still considered a southern school. Unlike Iowa State and other schools in the Midwest and North which had been accepting minorities for years, the southern schools had just started accepting Black students. Prior to that, you may have heard of George Wallace standing in the schoolhouse door in Alabama to keep Black people from the University of Alabama. This was only a few years after that, so the schools were just beginning to open up to minority students.
>
> Although Vanderbilt was a private school, the undergraduate school was still segregated in the early 1960s. We actually met one of the first Black undergraduates to be admitted to the undergraduate program at Vanderbilt. She was a friend of ours, and had preceded

> us by only a few years. That gives you some perspective on the time period in which we were matriculating.
>
> There were four of us in the graduate mathematics program at Vanderbilt at that time. That was a pretty big clique, so we could study together, and it didn't bother us that we weren't invited to be in the study groups with the other White students. We were able to study together and help each other. One of the four Black students was single and was lacking a social life, so she left after the first year. That left the three of us. We were able to gradually connect with the other students and eventually they received us very well. I didn't have any real complaints with the students there.

The Bozemans noted that living off campus was undoubtedly a big advantage to their social life. They heard from a fellow Black graduate student that "living in the dorm was so bad that the only people who would speak to you were the maids, and after one year, he had an ulcer and left [and got his PhD from a northern institution]." Sylvia Bozeman characterizes the social environment as "not exactly warm. We were breaking new ground, but eventually we made friends." Both of the Bozemans shared that they are still friends with some of their fellow Vanderbilt students to this day.

When they first began their studies at Vanderbilt, they heard rumors about a professor resigning his position in protest of Vanderbilt's decision to admit Black students:

> We heard the stories, but we had no way of verifying whether or not the stories were true. So we went into one of our classes and discovered that this particular professor who the rumors were about was our instructor.
>
> Q: What was that like?
>
> A: Interesting, you know. He was southern. He didn't show any signs of prejudice or anything. He just taught the class.

For Potter, the graduate school experiences of isolation reminded him of his early experiences desegregating his Baltimore high school:

> I keep reminding people that Maryland is called a *border state*. I can remember growing up, there were these coffee shops sort of like the

> McDonald's are today. They were all over Baltimore, they were specifically named The White Coffee Pot. What that literally meant was that if you weren't White, you didn't go in there. I remember being at Maryland on the campus. It took me back to Southern High School, [where] I was the only African American in my class . . . because when you are the only one, you don't see any. All you see is White people. . . . African Americans at that time were so rare on the campus. That is a big campus. I remember you would walk and if you spotted an African American, you would look across and go, "Hey, yeah, we're here!"

Johnson, the Bozemans, and Potter all found that their undergraduate education had well prepared them for graduate school. Succeeding at the doctorate is not just about mastering content and completing the dissertation—there is a culture of doctoral programs that also contributes to one's success or failure. Throughout mathematicians' narratives, the importance of networks and communities emerges as a central factor in their graduate school success—even when those networks and communities were not necessarily present within the graduate institution itself.

From their earliest experiences with mathematics, when family members, neighbors, teachers, and other school adults created spaces in which students learned and performed mathematics, to their college experiences with teachers, advisors, and mentors, who often ensured that they had opportunities to participate in high-level mathematics learning in and outside of college classrooms, Black mathematicians' love of the subject ensured that they remained captivated by its possibilities.

"It Was the Thing to Do"

Although most mathematicians in this study followed the traditional path of majoring in mathematics in college, then proceeding to graduate school in mathematics within a few years and completing the PhD, a few mathematicians in this study were not mathematics majors in college at all, entered the military, and/or worked for a number of years before pursuing a doctorate. One mathematician recounts that his going to graduate school in mathematics occurred relatively later than most. Having arrived at his military academy for college, he found that his high school preparation meant that he could skip the introductory calculus courses:

> So I started off in differential equations, [and] was taking class with all the upper class. At the military academy, the upper class don't just become buddies with the lower class. There is a military structure there, so I wasn't talking to many of my classmates outside of the classroom about mathematics.
>
> I did have another guy who was a math major on the team, and we would discuss homework occasionally, but not to the level of where we formed any type of regular discussion group or anything like that. I was really more isolated, you might say as a—call it a scholar-athlete who was a math major. I really didn't have those things that you're talking about. And that's why I say I bloomed late in life as a mathematician. It really didn't occur until, I guess, almost 15 years later when I was selected by the army to go back to graduate school. I was in these groups—we formed study groups to talk about math discussions, math tables—it was fascinating to me. It was a whole new world that opened up to me, being around a group of people talking about mathematics. So that came later, as a mature adult.

But he is an exception—most of the mathematicians in this study knew as high school or college students that they planned to pursue mathematics in graduate school. For some, as described in the previous chapter, graduate school itself was an unknown quantity. This was in part attributable to the fact that, as Leverett described in an earlier chapter, in the segregated South it was highly unlikely that a Black student knew any person with a PhD until he or she got to college. Even some of the third-generation mathematicians in this study stated that they lacked information about what it meant to go to graduate school in mathematics and what a career in mathematics might look like other than teaching high school. A third-generation mathematician who was a college athlete recalled an administrator encouraging him to consider graduate school:

> She was actually involved in the academic eligibility of the athletes, so when she had noticed that I had a perfect score on my AP calculus exam, she came to me and said you don't need study hall, you may want to think about grad school. I didn't really know what grad school was.

Other mathematicians had family members who had attended graduate school, some in the mathematical sciences or related areas. For some, it was a given very early on that they would attend graduate school:

> I always anticipated going to graduate school no matter what my major was. My dad used to say, "You go to college to learn how to learn, and you go to graduate school to actually learn something." So I always had the perception that I'd be going to graduate school immediately after, even when I was in high school, even though I didn't know I was going to be a math major. Yeah, it was the thing to do.

Ryan Kennedy, who recalled in an earlier chapter the "serious" mathematics books that were around the house, had a mother who had earned her master's degree in mathematics. Others had family members who had graduate degrees in statistics, psychology, physics, and other fields. When describing how he decided to go to graduate school in mathematics, one third-generation mathematician recalled:

> Well, I had an interest in math. [But] I think I would say that at least as motivating for me was—my dad had given me an article that year before I was about to graduate. I think there were four African American PhDs produced in the country. And not even all of those might have been in math proper. Some of them might have been in math education. And it was just, like, "Holy cow, are you kidding me?" And so that was certainly a motivating factor of "I want to do this." And, you know, try and increase the number in the pipeline, not only [for] me [to] do it, but [to] come back and get some other folks to do it, too.

For mathematicians who entered college knowing that they wanted to eventually earn the PhD and for those who were unaware of graduate school and career opportunities in mathematics, undergraduate faculty at their institutions as well as connections to mathematicians through summer research experiences or conferences further exposed them to what graduate school in mathematics and what careers in mathematics might entail. Many mathematicians speak of the importance of role models and mentorship by faculty. Undergraduate advisors and mathematics professors were often the ones who encouraged students to pursue mathematics and to go to graduate school. For some, this amounted to a brief conversation, but for others, this was engaged, prolonged encouragement, as one third-generation mathematician remembers:

> Probably during my sophomore year, I had one of those introduction to proof classes. Up until that point, of course, I thought that math was just a bunch of calculations, which I did like. But that is a misnomer, of course. So when I took that class, I just loved it. I loved doing proof. I just thought it was so magnificent. I thought that, "Maybe I don't want to teach high school. Maybe I can get into math." That is when I met my advisor, she taught that class. I think that she definitely saw something in me. I told her, "I am not sure that I want to teach high school." She began to tell me that I really had a knack for mathematics and started telling me about summer programs that I could attend. . . . I just would like to say that I would definitely have never gotten a PhD without her. [At first] I thought I would just get a master's mainly to buy some more time and figure out what I was going to do with this math degree. . . . So she definitely was the one who was like, "You need to apply to the PhD program." I am convinced that I would not have done it had she not been so proactive.

A second-generation mathematician shared a similar story:

> I think the person that I most wanted to be like would have been the only woman professor I had at [my undergraduate institution]. She is an important mathematician of the early 20th century. I think she probably would be a mentor. So going through it she would be the person who would have mentored me because I asked her some career questions. I had some real questions [about] whether I could become a professional mathematician. I'd only learned about the existence of that profession late in high school, maybe in college. And I wasn't sure I was the right social class because, you know, at that point it looked like a profession where, you know, these are typically people whose parents were professors, you know? And we lived in a housing project the first 16 years of my life, so we certainly weren't in that strata. I wasn't absolutely the top student in

> the department. I was certainly reasonable and I was respected as a talented student, but the question was whether I could go on and I remembered consulting with her. And she was quite encouraging, by the way. I think I still have a letter from her.

For some mathematicians, undergraduate faculty or even high school teachers steered them to their alma maters—for example, Clarence Stephens attended Michigan largely on the recommendation of his college instructor, who had gone there; Raymond Johnson attended Texas in part because his high school teacher had attended there, and then he attended Rice, as his undergraduate advisor was connected to that institution. When Sylvia Bozeman completed her PhD at Emory University, it was in part because of its convenient location, given her being an instructor at Spelman College, but it was also because Etta Zuber Falconer, her mentor and the department chairperson, had graduated from Emory herself and recommended it.

For most mathematicians, their undergraduate professors discussed their own graduate school experiences with them, provided or referred them to extracurricular mathematics experiences, and helped them navigate the graduate school application process. For some, undergraduate faculty continued to provide mentoring, advice, and guidance even when students had begun their graduate programs. This happened occasionally at predominantly White institutions, but such experiences were par for the course for Black mathematicians who attended historically Black colleges and universities (see chapter 5).

The Graduate School Experience

Staying or Leaving

Having, for the most part, succeeded in high school and college in multiple types of environments—small rural schools, large urban schools, predominantly Black and predominantly White settings, small liberal arts colleges, and large comprehensive universities—Black mathematicians found themselves making the transition (as one mathematician noted, "a social transition, a mathematical transition, a geographical transition") from college to graduate school. Attending public and private institutions such as the University of Illinois; the University of California, Berkeley; Howard University; Yale University; North Carolina State University; Vanderbilt University; Rice University;

Emory University; Massachusetts Institute of Technology (MIT); and the University of Michigan, Black mathematicians, like many graduate school enrollees, reported a variety of experiences in completing the doctorate in mathematics. Some reported that their experiences were relatively smooth, but some found the transition between undergraduate and graduate school to be difficult and had challenging periods in graduate school. These difficulties had little to do with mathematics content but were related instead to how the mathematicians adjusted to graduate school culture, norms, and hidden curricula (Herzig, 2004). Some were given a "heads-up" by their undergraduate advisor. As one third-generation mathematician recalled:

> I talked to my advisor about [choosing a graduate school]. That was the first introduction to how graduate school works and how important it was to kind of know the culture of a graduate department. As an undergrad, before that point, I didn't know. I just thought it was kind of like undergrad. I didn't realize that it is really, really different. My advisor definitely went through issues at [her graduate institution] that she shared with me about not feeling welcome and that type of thing that goes on. I had not known that, especially being in a very protective [HBCU] undergraduate institution. You would think that everyone is like that. So she talked to me about the importance of kind of knowing the culture of the department that I was going to apply to.

Although this transition is difficult for many regardless of racial background, it can be particularly difficult for those who are underrepresented in graduate mathematics—women and members of ethnic minority groups (Case & Leggett, 2005; Herzig, 2004; Murray, 2000). If their ability in mathematics had not been challenged before—if it had not been challenged in high school or undergraduate studies—graduate school could become a kind of crucible for some mathematicians. This is undoubtedly true for many doctoral students, but for Black students, the overlay of race and perceptions of ability are vividly present in their narratives about these experiences. For Black women, particularly those who attended HBCUs, it was often the first time that they were confronted with the "double knot" of race and gender, interacting with professors and fellow students who had low expectations of their mathematics ability because of their gender as well as those who had low expectations of their mathematics ability because of their ethnicity. No Black women in this study, regardless of generation, mentioned feeling isolated in mathematics in her high school (whether she had attended a predominantly White or predominantly Black high school), and very few reported feelings of isolation in their undergraduate experiences. As Francine Robinson noted:

> Nobody ever made it seem like math and science should be hard, and so I think I just ended up doing well in it, because I didn't know it was supposed to be difficult or had to be difficult. But I think some people feel like it's just supposed to be hard or something, and I just never felt that way.

Of the 10 women mathematicians who participated in this study, 6 attended historically Black colleges and universities. Women who attended these institutions describe no instances in which expectations of them seemed low related to their gender (I discuss these HBCU experiences more fully in chapter 5). One, involved in a graduate mentoring program for women, noted:

> The women I know that went to big majority schools and try to be a math major get crushed. So the women in graduate programs tend to come from smaller schools. So forget about race, forget about gender, so maybe that's how they don't fit the successful graduate school package. Or coming from the South, you know, because people have all kinds of judgments about that. I mean, it's just—everything plays into it.

Now it is true that this is a select group of women who persisted in mathematics through the PhD, and so it is quite possible that Black women experience more acts of racism and/or sexism that discourage them from pursuing mathematics than are recounted in this book. But for women participants in this study, it is in graduate school that Black women seem to experience egregious acts of sexism and racism.

Mathematics is still a field in which the common trope about women in mathematics is that women cannot do mathematics at a high level, exacerbated by commonly recounted prejudices of people in the popular press, professors, and even a university president (the 2005 comments of Lawrence Summers, president of Harvard, come to mind). Less common, one hopes, nowadays, is the supposition that women don't "need" PhDs because their husbands will have one. As one mathematician recounted:

> The graduate chair called and gave me this nice, long conversation. He told me I didn't need a PhD. . . . That my husband's going to have one and it would be too hard for both of us to get a job if we both had a PhD. . . . I just dismissed him. I thought that was outrageous but I did not say anything there.

There are still instances in which Black women mathematicians feel that their gender is a problem for their colleagues and faculty at their graduate institutions. Black women in this study were more likely than men to mention the "graduate school switch"—earning the master's at one institution, then finishing the PhD at another. All who mentioned this—male and female—pointed to acts of racism, either at the institution itself or within the department.

About her initial graduate school experience one mathematician noted:

> [That place] was traumatic. My self-esteem took a beating. It was a very, very difficult time. The department was mostly comprised of White men who first of all did not feel like women could do math, let alone Black women, let alone a Black woman from a Black college. I just had too many strikes against me. . . . That was one of the worst times of my life.

Her next graduate school experience was much more positive, largely due to the efforts of the department chair at the time:

> The department was bigger, so there was more [faculty diversity]. He [the department chair] made a concerted effort to really recruit Black students to the program, so there was a big network of Black students there at all levels so they could help each other as they went through the process. . . . Even with the White faculty it was a different feel. I didn't get that same sense of lack of belief in me. I think that grad school, in their minds, was a weeding-out process for everybody, not just a select group of people. At [her first graduate institution], I felt like we were singled out. At [the second], they were like, "No. Grad school is hard. You've got to work hard." I respected that and I planned to work hard. Just give me the chance to work hard. It was a different feel. It was a much better environment.

Women are still a significant minority in graduate programs in mathematics and on mathematics department faculties.

> My graduate program was a culture shock. So, I was like the only woman and African American pretty much. In the first-year class there were like nine of us. There were three Hispanic guys, there was a Russian guy, two White guys, a couple Asians. I mean it was a very diverse group. But yeah, it was definitely different. I remember I would get called out when I would miss class because it was so noticeable. I mean it was just so obvious when I wasn't there because I was the only female and the only Black person. You know? I did feel like I stood out a lot more. It was just easy to notice if I was somewhere or not. The faculty were very supportive. I never felt like they didn't want me to succeed or anything. There was a Black woman who had graduated the summer that I started. And so I actually took her desk, took her office spot. But yeah, they were like very supportive of African Americans. But it was just different. It was definitely a culture shock.

The feeling of being "singled out" relates to Claude Steele's description and analysis of how individuals (regardless of demographic background) respond to a perceived "spotlight" effect when they participate in activities where they are in the minority, and that can contribute to depressed demonstration of one's academic talent (Steele, 1997). Steele calls this situational effect, when one has the potential to confirm a negative stereotype about one's group, "stereotype threat," and it has been documented to have an impact on groups as varied as women in mathematics and African American college students. The impact of stereotype threat is most present for people who are strongly identified with the domain in question—that is, for example, the impact of stereotype threat on mathematics performance is most significant for women who consider themselves "math people."

The most common issue reported by mathematicians as it relates to doctoral studies is being isolated in doctoral programs. For a number of Black mathematicians, whether or not they felt isolated or unwelcome were significant experiences that influenced their choice of graduate school and whether or not they stayed in or left PhD programs. Said a third-generation mathematician, Laverne Richardson:

> Everyone, no matter what the race, like the graduate chair, everyone was very. . . . They really wanted me there. I felt like it wasn't just, "We need to admit somebody." I felt like they really, really wanted me there. They talked to me about everything very, very well. The

> graduate students were honest. . . . I talked to them by myself, so I am pretty sure that they would have told me the real situation. They just were very, very honest in saying about how people were supportive. I think that sold me even though it's more of an applied department. They definitely have pure mathematicians, but despite that fact, I think that the culture kind of outweighed the "what am I going to do?" for me. For some people, they would never go to a department if it did not specialize in the area in which they wanted to specialize.

One mathematician had been admitted to one graduate school for the PhD program but ended up at another after "feeling no connection" to the first institution and, further, feeling "no attempt to connect to me." Running into a math professor from Howard University on a trip to Washington, DC, she was encouraged to consider going to Howard instead.

> What I needed at that time was feeling welcomed. I think if my undergraduate experience hadn't been so alienating I would have been a lot more discriminating in my choice and I would have made a more informed choice, but that professor being there and being so friendly and welcoming and seeing all these brown people. . . . I was just like, "that's the place for me."

It was one thing to be the first Black student at a White institution in the 1940s, 1950s, or 1960s—it was another to be the first at a White institution in the 1980s or 1990s. Yet Black mathematicians still blaze these kinds of trails. Even if they are not the first Black student to enroll in a graduate mathematics program, they are often the only Black student in graduate school mathematics courses. Although some are able to connect with their fellow students, others find this to be a challenge, and it is one reason why some Black mathematicians leave their graduate programs. As one third-generation mathematician reported:

> It wasn't that the other students didn't allow me into the group or it wasn't that I couldn't go; it is just that I stayed to myself and they stayed. . . . Because they formed groups there. I guess I just wasn't in any group and I found that I was very, very isolated. I was in a four-story building in the science complex and I think there was one

> other African American, which was the cleaning lady. Then there was a guy from one of the African countries. Other than that, we were the only Blacks in the whole building . . . [among] the graduate school *and* on the staff. I struggled through that and I left before I finished.

Thus, these experiences of isolation resonate not only with those who were among the first to desegregate their graduate institutions in the 1960s and 1970s but also with some mathematicians who entered graduate institutions in the 1990s and 2000s. One third-generation mathematician who entered a program largely on the strength of its relatively diverse faculty noted:

> But that department itself wasn't prepared to deal with Black graduate students. So, you know, afterwards, I don't know if I was the first, but there was a slew of Black graduate students who went through the department, left, went on to other departments and did very well. And I think it's all sort of attributed to this one guy who would say these horrible things about not being able to succeed in mathematics, pull you into his office, like, "You shouldn't be here," and stuff like that. So that turned out to be a disaster, but I was there for three years trying to get a PhD so I ended up leaving with a master's.

Another third-generation Black mathematician who started his graduate school career at one institution left, not because he did not like his department, but because of the virulent racism on campus and in the surrounding town. He described the town as "so racist—I mean, God, I was called nigger so often," and episodes on campus (a caricature drawing of a Black man being hung was posted all over the campus) and off (the Ku Klux Klan held a rally not far from the campus) contributed to his decision to leave:

> I liked the department, but I hated living [there]. . . . So after I finished two years I wanted out. My plan initially was to stay there and get a PhD, and I probably would have if I had felt comfortable walking home, but I was still uncomfortable walking home. And I just didn't like it. I liked the department, I have to admit. [My advisor] and some of the other faculty there I thought were really good and were very protective of me and that sort of thing.

Even when graduate programs on the surface appear to be diverse in terms of students and/or faculty, mathematicians report that "diversity alone is not enough." As these stories demonstrate, "inclusiveness and equity" must also be present within doctoral programs (Falconer, 1996, p. 169).

Navigating the Graduate School Terrain

In choosing to remain at institutions, some Black mathematicians describe the situations they face as being par for the course for [Black] graduate students in general:

> Being in grad school is a very personal process, you know, like you have to give up so much of yourself, not just for the work but because they want to see if you're okay. It takes part of you. In order for you to do well with it, you have to be motivated. If you dislike the people you're around, you're going to be less likely to [succeed], and I think it's especially difficult being a Black person in this country going to graduate school, you know, the isolation and all the rest of it, and the commitment . . . then to be isolated and have to be committed to people who make you feel funny.

This larger backdrop to the issues facing graduate students in general was a theme shared by mathematicians who had been graduate students in small and large graduate programs, northern and southern institutions, and public and private universities. As the mathematicians in the previous section stated, the culture of a graduate school mathematics program is critical to student success, even if, as in at least one case, a strong and welcoming department was not enough to make up for the racism a Black mathematician experienced on campus. What makes a "good" culture in general and for Black mathematicians in particular? In many cases, predominantly White institutions that have graduated "significant" numbers of Black PhDs in the sciences have committed faculty and/or administrators and/or strong alumni networks of Black graduates who share information about how to navigate the institution. Some of these institutions (Michigan and Maryland are examples) have a historical legacy, if not a reputation, for being receptive to having Black mathematics graduate students, and this is often generational in that one sees Blacks earning the PhD in mathematics from these institutions across multiple decades. In addition, Howard University has graduated a significant number of African American mathematicians since the inception of its doctoral degree program in 1975. One faculty member there, Neil Hindman, is responsible for sponsoring the largest number of Black students who have earned their PhDs in mathematics. In interviews with Black mathematicians, the role of faculty in crafting graduate school environments conducive to Black students' success emerged as key.

The Role of Faculty

As in undergraduate institutions, faculty are critically important to Black mathematicians' acclimation to and integration within graduate programs. However, it should be noted that several mathematicians mentioned professors (especially Black faculty) who were integral to their success from outside of their institutions—their own undergraduate faculty, as well as faculty with whom they came in contact as graduate students at various conferences. When some Black mathematicians speak of influential faculty from their graduate programs, they mention faculty who may or may not be their primary advisors and describe instances in which they chose advisors based not just on the type of work that their prospective advisors were engaged in but also on the basis of the personal connection that they had with those advisors. Some found that they had to develop additional areas of expertise to work with those advisors, but this was a price they were willing to pay. Robert Bozeman, who earlier mentioned knowing the first Black undergraduate to desegregate his graduate institution, remembers one faculty member in particular very fondly:

> Billy Bryant was a professor at Vanderbilt and a very friendly, outgoing man. He went out of his way to try to make us comfortable. He was always saying, "How are things going? Come by my office and see me." He was that kind of guy. I actually ended up writing my master's thesis—he was a topologist—with him. At that time, we were required to write a master's paper.
>
> When Dr. Bryant retired, many students wrote congratulatory letters, including me. I expressed in my letter my appreciation and admiration for him. I told him that the reason I wrote [the master's thesis] with him was because he was the only one that was so friendly to me. In one of our subsequent conversations, he told me that he hadn't realized the impact that he had on me. I said, "You were so friendly, that when it came down to it, there was no one else with whom I wanted to write my paper." There were other nice people in

> the department, but they didn't choose to talk with me like he did. The only exception was my PhD thesis advisor, but he came later. He wasn't there my first year. I worked with Billy Bryant and he encouraged me to stay on for the PhD.

Younger mathematicians also mention the importance of personal relationships with faculty in terms of their "making it through" graduate school. One third-generation mathematician, describing his discomfort with an advisor who made presumptive racial comments and allowed students' prejudicial commentary in seminars to stand without comment, began working with another faculty member in an area in which he was less prepared.

> He pushed me a lot in that area, got me a lot better at that stuff. . . . He basically made me take a qualifying exam just to work with him. So I spent a year just studying in preparation for this exam with him. Then he was just this sort of person to just have personal conversation with you, like, "How are you doing? Are you exercising, feel well?"—things like that. And he would tell you his story, like how long it took him to finish grad school. He was very personable.

Noteworthy about these kinds of relationships is that they include strong mathematics teaching and learning, but there is also a personal dimension that Black mathematicians value. There is acknowledgement that there is more to the PhD process than just learning the content and writing a dissertation. In talking about advisors and close relationships with other faculty members, Black mathematicians do describe relationships that are sometimes grounded in mutual mathematical interests, but they frequently highlight the importance of shared personal connections that are not necessarily based on race or gender. As one mathematician noted, her mentors have been male and female, Black and White. Another said that his advisor was "a brilliant mathematician, but what I saw in him was a good person." These authentic relationships serve to contribute to Black mathematicians' being successful at navigating the sometimes rocky graduate school terrain.

The (Hidden) Curriculum of Graduate Programs

The graduate school process—course work, qualifying examinations, dissertation proposal (and finding an advisor), and dissertation—is fraught with layers of tangible and intangible benchmarks. Phrases such as "weeding out," "getting stuck," and "going in circles" abound in the literature about the process of becoming a mathematician, and they certainly appear in these mathematicians' narratives. Even at the beginning, choosing courses to take—upper-level undergraduate courses or graduate courses—is a decision that can have significant consequences. As one mathematician noted, it is important for one's identity as a graduate school student to take graduate mathematics courses, but there may be courses in which a strong foundation is required. For mathematicians who attended small colleges with sometimes limited course offerings, taking a graduate-level topology or complex analysis course in graduate school might lead to difficulty. Another popular theme in mathematicians' narratives is the importance of a strong undergraduate foundation in proof; as one mathematician recalled, "the first year of graduate school I saw students drop out, from all sorts of schools. I really felt that it was because they didn't have that core foundation of proof," and the transition to courses with proof was extremely difficult. Most of the mathematicians who reported having difficulty in a course attributed that to weaker preparation in a particular content area in their undergraduate institutions—this was particularly the case for those who had attended smaller institutions. An exception was the experience of Arthur Grainger, Earl Barnes, and Scott Williams, who were part of Clarence Stephens's student study groups at Morgan State their first year and continued in the program that he started. As Grainger recalls:

> My first class [in graduate school] was analysis. The professor walked in and looked and said, "No. This won't do; some of you have to leave." Everybody looked around. He said, "Okay. This is what this course assumes." He went and started writing. . . . Have you ever seen those cartoons about those professors writing with chalk in one hand and an eraser in the other? . . . The board is divided into three sections. He would go and write here and when he would get to this section, he would start erasing and writing and erasing. The thing is that he could write on the board faster than I could take notes. So I had to speed up. I was looking at what he was doing and just in that first hour he covered everything that I saw in advanced calculus at Morgan. Then after that hour, he said, "This is the stuff that you have to know for this course. If you [do not know it], perhaps you can take [another course]." I would say that about two-thirds of the class rushed up to sign to get out of the class because they realized

> that they didn't have the background. They hadn't seen that stuff before. So the next day the professor came in and there were about 15 of us and he said, "That is better."

Qualifying examinations are a significant hurdle, the key determinant as to whether or not mathematicians can continue for the PhD or earn the master's degree and an invitation to leave the doctoral program. (Some women mathematicians, however, report instances in which they were encouraged to leave despite passing qualifying exams, because "no one" was interested in working with them.) Having failed part of his qualifying examination, one mathematician who earned his PhD in the 1970s describes his decision to remain at the institution:

> They had a policy that you could take your qualifiers twice, but if you failed them the second time they would stamp on your record that you were expelled for the lack of performance on your qualifier. And so a lot of my White colleagues who did not pass all of their qualifiers the first time, they went to other schools. So, they didn't want to run the risk of having anything on their record about them not being able to pass qualifiers and not being able to do any further study at [the institution]. They thought it would affect them with other schools. . . . And some of them asked me wasn't I going somewhere. I told them, "Hey, it's going to be the same thing for me anywhere I go." At that time there were no predominantly Black schools offering PhDs in mathematics. And the climate was about the same. The majority of the schools in the nation had never awarded a PhD for Black folks, the majority of doctoral institutes. So, I told them I had just as good a chance [there] as I had in any. I stayed on there and took my qualifiers in those two subjects a second time and I passed.

A third-generation mathematician who passed his examinations on the first attempt described the potential stress of not passing, reflecting on other Black graduate students' experiences and what failure would have meant for his HBCU alma mater:

> This is [a situation] that affects us, I think, more. Now it's stressful on anyone, anyone who has a goal, sure, you want to succeed and so the thought of getting put out of a program because you couldn't pass an exam, that's stressful. But I think it's especially stressful for

> [Black students], and I've seen it, and I've lived it with friends, both in my department—not only in my department, other disciplines, other universities, the same thing. That, you know, this [failure] would have reflected on Black folks. *The extent to which we're really representative, you know, this sort of burden that White folks don't really have.* [emphasis added] It would have just killed me not to have passed. I've seen it through the years with our alums who go places if it didn't work out, knowing what they go through because I talk to them. And then friends in other departments as well, a friend of mine from [another HBCU] was [at the same university] and again, it just devastated her. And because we were in a similar boat, we could talk about these sort of things, about what it meant that she felt like she was letting down her whole department back home because she'd been their star, and the pride they see sending us off [to these institutions], and it's a burden.

The feeling that individual Black students have that they are considered to be representative of entire demographic groups is supported by the experience of another third-generation mathematician, who applied to two different graduate programs with the same university, heard from and was accepted by one, and was ignored by the other. When she later arrived at the institution, she asked a faculty member about her application, saying that she'd been disappointed not to hear from them at all:

> And he was like, "Well, you know, you looked really great on paper and everybody was really impressed. But, we had a student who came from [your alma mater] that left with a master's. And so they figured that maybe you wouldn't be able to do the program." And I was just like, "Are you serious?" I said, "Well, have you guys ever had a person from Berkeley leave the program?" Because I had known that they had these guys that came from Berkeley. And he was like, "Yeah." And I was like, "So do you no longer accept these UC Berkeley students because [of] this person?" He was like, "Well, no. Of course not." But it was just kind of like how could you, this one random student who came and left your program? And how could you judge my application based on this person? And I just thought that was crazy. You know, that's sort of how they did. It wasn't just HBCUs. Even the smaller schools, like small state schools, you know they had more students who did well, then they'd be like, "Oh, well

that school must really prepare you." Or they had some one student who did poorly, now that school's no good. It was just sort of sad the way that they would screen students, not necessarily based on their qualifications but based on what school they came from. "Oh, that's a Berkeley student. They must be good." Even though they could go there and do poorly. It was like, "Well, they come from Berkeley, so we'll give them a chance." I thought that was unfortunate. In hindsight, it was better that I didn't go [to that program].

Kinships and Communities in Graduate School

Student communities in graduate school are important facets of the graduate school process and, for many mathematicians, are integral to their success as doctoral students. These communities range from student study groups and seminars to Black graduate student associations within institutions and later expand to student networks that are facilitated by professional organizations. Mathematicians in this study have varied beliefs about the importance of study groups to the graduate school experience—some acknowledge that studying with groups for qualifying exams and courses was helpful; others did not participate in such groups. Some note that they were never asked to belong to a group. As one second-generation mathematician recounts:

> There were study groups at [the institution] for the different classes I was taking. But my classmates told me that these groups didn't exist and told me they were struggling on their own. But that was because of the fact in almost all of my classes, if not all of them, I was the only Black and I just wasn't invited or welcome to those study groups. And so I had to do it on my own.

Others find that on occasion working with other students can raise uncomfortable and unsettling racial issues. A third-generation mathematician describes his experience in a seminar:

> I remember them having a conversation about some model they had to predict how many people you knew in prison. So we're talking about that, and they were talking about Black people and how they poll, they knew everything. I remember sitting here the whole time. And then they went on to talk about how Black people should take

> a different SAT because the SAT didn't measure them correctly. So I suppose they felt they were being enlightened and things like that, and I'm, you know, it was upsetting. And I'm sitting there thinking, "I probably did better on the SAT than everybody in this room."

However, for several mathematicians, collaboration was an extremely positive experience. They mentioned the importance of faculty or fellow students who were not their advisors in terms of their induction into graduate programs. This tradition, which in some ways parallels the experiences of David Blackwell (whose fraternity brother told him about the courses needed for graduate study in mathematics) and Clarence Stephens (whose undergraduate advisor steered him to Michigan for graduate school because that was the school the advisor had attended), is shared by several mathematicians of the second and third generations:

> First of all I should say about the [graduate school] experience one very special thing happened. . . . When I was accepted it turned out that the big guns in my field were not going to be there in the fall. However, there was a postdoc there who was Black. We started working together the spring before I was formally enrolled. And he was a major mentor to me. . . . In all aspects he looked out for me, even after he left, and so he was very important to me. [After I arrived] he was there I think another two years, and he supervised my oral exam, for example. That made a terrific difference. Absolutely.
>
> Q: What do you think it would have been like if you hadn't met him?
>
> A: Much more difficult. I mean sort of all aspects—I mean he kind of guided me through everything, always supportive. And then after he left he sort of passed me off to other people. He sort of found somebody else to look after me. I had a decent advisor, but not as close. I mean my advisor was friendly, but the point is I sort of had all the extra help, extra resources.

In short, we see that the theme of kinships continues to be an important component of graduate school for Black mathematicians. Individuals and formal and informal student communities play a role in supporting Black mathematicians' learning and success in graduate school and inducting mathematicians into the profession.

> Even though there was a good group of us, we did not limit ourselves to just us in terms of studying. We would study with other people in our classes. I do remember when we were studying for the qualifying exams, we were pretty rigorous. We were like, "We're meeting in the library at 8:00 a.m., and we'll be here. We'll be studying until 5:00 p.m." We did do that, I remember. I did take a picture of it. We would be all around the table studying for those qualifying exams.
>
> That was a great experience because everybody was supportive of everybody. You did not feel bad for something you didn't know because you knew that somebody else understood and could explain it to you. We all had our different strengths and we brought our strengths to the table. In that respect, I thought we had very effective groups.
>
> What was cool about us was that we were part of the BGSA [Black Graduate Student Association], so we would do all kinds of things together. With math students, when anybody achieves something or passes qualifying exams or oral exams, we would all go out and celebrate in honor of that person. At BGSA, we would have potlucks or we would go to movies together or do community service together.

These groups at some institutions also incorporated social activities, and faculty on occasion supported those efforts. One doctoral advisor held seminars on Friday evenings that incorporated a social hour as well as lectures and study.

> I thought as a student that it was lectures and on Friday night at 5:00 I did not want to go to a math lecture. I wanted to go home. But actually that became a big thing for me and some other students. We all started going to this seminar every Friday and it was a big hangout. We'd go to dinner then go to the seminar and stay, until, God, sometimes we'd stay there until 12:00 [midnight], you know? It was really late. We'd stay late. Sometimes [our professor] would stay with us. He'd do stuff sometimes like at 10:00, you know, he'd go out and bring snacks too. And we'd sit around and talk some more, you know? So we'd stay up. That actually was one of the most fun times of my life, to tell you the truth. . . . Yeah. We had a good time and I was doing mathematics.

Sometimes Black students took the initiative and deliberately created their own groups. As one mathematician describes, she thought it unlikely that anyone would reach out to her to participate in a group:

> I think you also need a group of students to work with [to succeed in any graduate program]. So that first year I sent out this email and said that we were going to meet on Wednesday for lunch and then we were gonna work on our homework together in the library. And I just sent it to all the first-year students. And people were like, "oh, okay." I mean like they were cool. Everybody showed up. We did that every Wednesday. . . . And it really helped us to grow, but the only reason I did it was because I was like "nobody is just gonna reach out to me to form a study group unless I form one myself." . . . It was so not official. It was just me deciding that I would rather we all work together so we could all learn from each other.

Professional Organizations: Bridges to the Profession

Another point of entry for Black mathematicians' induction into the profession is via their participation in professional organizations. Mathematicians in this study participate actively in the oldest mathematics associations in the United States, the Mathematical Association of America and the American Mathematical Society, as well as in other professional organizations, presenting papers, giving lectures, and serving on committees both as graduate students and after they earn their PhDs. Several mathematicians who participated in this study have been recognized with awards by these organizations. However, Black mathematicians have not always been welcomed with open arms.

The treatment of Black mathematicians by professional organizations during the early to mid-20th century is a shameful story recounted in *Black Mathematicians and Their Works*, published in 1980. At best, these mathematicians were considered invisible; at worst, they were humiliated and endangered. In 1951, Lee Lorch, a White faculty member in mathematics at Fisk University, a historically Black university, wrote a series of letters to the leadership of the Mathematical Association of America challenging them to actively, rather than passively, repudiate racist practices by subsidiary chapters:

> The southern meetings seem to have been organized around the assumption that no Negroes will attend. The arrangements

> committee for the Association's southeastern regional meeting held last spring at Vanderbilt and Peabody listed only housing facilities restricted to white patrons. . . . One Negro mathematician did attend [another meeting in Alabama]. . . . The program listed a Social Hour, details to be announced at the meeting. He asked at the registration desk for further information. A member of the Arrangements Committee told him that "technically" he could attend, but that he "probably would not want to do so, as it was being held in one of the girls' dormitories." (Newell et al., 1980, p. 316)

It is clear that such practices by professional organizations were felt deeply by Black mathematicians of a certain generation. Dr. Mae Pullins Claytor, writing to the editors of a collection of mathematical papers by Black mathematicians about her deceased husband William Claytor's (PhD 1933)[2] work, stated:

> There is so much I just cannot put on paper. . . . I thought about the days Bill used to tell me how owing to the Black-White mess, he had to stay at a private home when the others were at the hotel where the [Mathematical] Association [of America] met. Over the years when the color-line became less, he never would attend any more meeting[s]. Kline [Claytor's adviser from the University of Pennsylvania] used to come to see us periodically and try to get Bill to go with him but I guess the hurt went too deeply with him. (Newell et al., 1980, p. 321)

These sentiments were echoed by White and Black colleagues of Claytor, who shared their memories of him in the *Proceedings of the Eleventh Annual Meeting of the National Association of Mathematicians* (Donaldson, 1989). As James Donaldson (himself a Black mathematician and dean at Howard University) writes: "it is now acknowledged that the racial practices in this country had a negative impact on the full contribution of his mathematical research genius and that mathematics and science are the poorer by far because Black scientists of Claytor's caliber were not given the opportunity to contribute according to their talent" (p. 454).

Due to these kinds of discriminatory practices and their belief that key elements of professional development were missing for younger generations of Black scientists, Black mathematicians (and other scientists) created their own professional organizations. In 1969, a group of Black mathematicians founded the National Association of Mathematicians (NAM), along with several smaller regional groups that were affiliated with NAM, primarily to provide opportunities for Blacks in the mathematical sciences to come together, discuss their research, and present papers (Houston, 1999). At Bell Laboratories, Black scientists and others founded the Cooperative Research Fellowship Program in 1972, a program through which a number of Black undergraduates in mathematics have gone on to become PhDs in the sciences. Unfortunately, this program is now defunct. In 1995, William Massey, along with several other Black mathematicians, formed the Conference for African American Researchers in the Mathematical Sciences (CAARMS) (Riordan, 2006). These mathematics organizations have their origins in older networks formed by Blacks in the sciences, including Beta Kappa Chi (Donaldson & Fleming, 2000), founded in 1923 at Lincoln University because Blacks were largely unwelcome in national mathematics and science organizations before the Civil Rights era.

Within and beyond the activities and annual meetings of these organizations, many Black mathematicians create or belong to informal and formal networks that support their own work and others' induction into the field. Many Black mathematicians informally mentor students or, more formally, structure summer research programs to which graduate students are invited. For example, two mathematicians in this study participated in a summer research program that David Blackwell hosted at Berkeley in the 1970s. Black mathematicians have formed such communities to facilitate access for others and their own socialization into a profession that has not always supported their talent and development. William Massey, one of the founders of CAARMS, in describing his work recruiting Blacks into mathematics careers, noted that "rather than waiting around to see if I would be getting new colleagues, I have been busy creating them" (Kenschaft, 2005, p. 194), a sentiment echoed by a number of Black mathematicians.

Many Black mathematicians, finding themselves isolated in their institutions, participate in professional organizations for the explicit purpose of meeting and, in some cases, recruiting other Black mathematicians. As Harold Prince noted:

> I definitely want to leave somebody behind [at my institution]; that's an important thing to me at the moment. I want to figure out how [it] is that I can retire and have another Black mathematician here, even another Black scientist. I think I am probably the only Black here in science. . . . That's really disturbing to me. I don't know exactly

> how to change that, but I definitely want to leave something behind. That's one of the reasons I'm interested in [these organizations]; it helps me find out who's available, so that is very important to me.

Prince, who earned his PhD in the 1970s, did not attend, nor does he work at, an HBCU. He notes that HBCUs have a built-in network that mathematicians who did not attend those schools lack, but for organizations like CAARMS and NAM. He notes "[these organizations are] very helpful because we are able to try to make connections and share information."

Mathematicians who attended meetings of NAM and CAARMS as undergraduate and graduate students attest to their importance and success in facilitating connections and conversations between students and faculty. Sample activities at CAARMS and NAM include student poster sessions, mini-courses, award lectures, and multiple opportunities for informal conversation and networking at receptions and meals. In addition, proceedings of these organizations' conferences are published regularly and include mathematical research papers by faculty and graduate students and, quite often, historical papers. In the last 20 years, historical papers about varied topics, including the founding of NAM and the life and professional experiences of J. Ernest Wilkins; surveys of Black mathematicians; and reflections by Lee Lorch, Etta Falconer, and others, have appeared in these proceedings. In his address to the second CAARMS meeting in 1996, published in the proceedings that year, Lee Lorch noted the "inspirational" quality of the conference:

> It is more diverse than the mainstream mathematics gatherings. We have young and old, junior and senior faculty, undergraduate and graduate students and we are all talking openly to one another, learning from one another, developing friendships, establishing badly needed networks, exchanging experiences, laying the basis for mentoring at all levels. . . . The gender distribution is more equitable, the "racial mix" more of a mix . . . these Conferences [are more] sharply focused on the African American community than . . . supposedly "general" meetings, another proof of the need for these Conferences. (Lorch, 1996, p. 158)

Mathematicians who attended these conferences as undergraduate students have shared remembrances of meeting people who were mathematicians and learning just what could be done with their mathematics degrees; they also received encouragement and advice about graduate programs. One third-generation mathematician attests:

> My mentors were Black mathematicians. It was good to see other Black mathematicians and to get advice from them as to what it meant to be a mathematician and how to be a mathematician. I met Dr. Johnson, who was the chair of the department at Maryland, at this thing called Math Fest. They have it every fall, where they bring undergrad students together and they make poster board presentations and talk about grad school. They have a grad school panel. I met him there and he is also one of my mentors. Sylvia Bozeman I met at a Math Fest and she became one of my mentors.
>
> The thing is is that our community is so small that we go to [NAM's] Math Fest or this other conference called CAARMS—Conference of African American Researchers in the Mathematical Sciences—I would see other Black mathematicians there. I would just sit down and talk to them and they would give me advice. I remember I had brought my real analysis book to one of the math fests and Dr. Johnson was tutoring me. Then I was emailing him, "Help," and he would email me answers. That is how they would mentor. They would tell me about what the life of a mathematician is. They would talk about graduate school and how to get through graduate school and what to think about it.

Even after receiving the doctorate, mathematicians continued to receive important professional advice from senior people in the field. As one third-generation mathematician describes:

> My fourth year of my [first] tenure-track job, [a senior faculty member] and I were having dinner at a CAARMS conference. I was talking about tenure and he said, "You know, if you are happy where you are that is fine, but if you think you might not be happy, you might start putting the feelers out during your fourth or fifth year for tenure." I did that to see where I was and whether I could move to a different place because I really wasn't happy at [my first institution]. Through my putting the feelers out, I got a couple of calls and ended up here, which I love.

An organization that focuses on the recruitment and retention of diverse women in mathematics, EDGE (Enhancing Diversity in Graduate Education), was founded by Sylvia Bozeman, herself a Black mathematician at Spelman College, and her colleague at Bryn Mawr College, Rhonda Hughes. EDGE brings small cohorts of diverse women entering graduate school together in the summers and provides mentoring (by EDGE alumni and faculty) throughout the academic year and the women's graduate school careers. Several mathematicians in this study have worked with EDGE, either participating themselves when they were graduate students as mentors and mentees or teaching during the EDGE programming, as some male mathematicians do. As one alumna of EDGE describes:

> That EDGE program is priceless. I cannot say enough positive things about what that program does and continues to do even with the cohort that we built. As we progressed, we still are in contact with each other. So it is helping even now at this level in my life.

Becoming a Professional Mathematician

Even in the 21st century, race is an ever-present specter in the professional lives of Blacks in the sciences:

> A new administrator told a young Black faculty member who was an alumnus of the institution that there were remedial programs there that were designed to help minority students who struggled with mathematics and science courses. Further, he was sure that the Black faculty member had benefited from such programs during his undergraduate experience.

Such incidents, recounted by Black mathematicians of all ages, indicate the still somewhat "anomalous social and intellectual position" of Black scholars in the United States (Winston, 1971, p. 678). These assumptions and discourses about intellectual ability and merit are rooted in historic and pervasive narratives about Black achievement that are focused on underachievement (Gutierrez, 2008; Martin, 2009a, 2009b; Perry, 2003). These narratives infiltrate discussions about Black mathematics achievement all along the mathematics pipeline: for example, the narrative of underachievement in mathematics is perhaps one reason why so many Black mathematicians report, despite their having excellent grades or test scores in mathematics, that the intervention of their parents was needed to ensure that they were enrolled in advanced mathematics classes. It is perhaps one reason why Black colleges are largely overlooked in broader discussions of mathematics excellence (to be discussed in chapter 6). Coded commentary about "motivation," "effort," and "engagement" often masks an undercurrent of low expectations for Black achievement in mathematics. One mathematician, early in his mathematics career as a postdoctoral fellow, relates a conversation in which a senior Black colleague admonished him about the language he was using to categorize his own work:

> Bob was asking me how things were going or something, so I made some comment about things and I wasn't getting a lot done. I said something about being a little lazy or something. He lit into me. Lit into me. He was so right. He said, "I'm going to pass this along to you," he said, "Don't do that, White folks are quick to ascribe the adjective *lazy* to us." He said that he sees how hard I work around there. So he said lazy was not the word for me. Yeah, sure, I was pouring my hard work into those students around there, but I was working hard. And yeah, I wasn't getting the job done research-wise and I needed to readjust some things. "But you're not lazy." And that was very big.

As professors, Black professors are often in the minority on campus or in their departments. The same spotlight effect they felt in graduate school often comes to bear when they are faculty members. Being on the tenure track is a difficult proposition for many junior faculty, and mathematicians reflecting on their experiences describe difficulty navigating politics and factions within and across departments, as well as struggles adjusting to the pace of academic life and competing demands for service, teaching, and research. These issues are not specific to Black mathematicians, but at times navigating them might be complicated by issues of race. One mathematician noted that she felt compelled to dress in a particular way:

> I always try to dress nice. Like I'm not going to wear jeans and a t-shirt. . . . I've been waiting to get a job so I could not dress like this. I'm just like, decked out every day. But I think the [students] really feel like I take it seriously. Some of my colleagues, [students] can go in there and they chill. And I'm like, "don't come in my office

> chilling." . . . Yeah, I think the whole image [is important]. And I know a lot of students weren't used to being taught by a Black woman. And that's the other reason sort of to really command that respect. Like, you know, I'm not on your level. I'm not your peer. You can't look at me like I'm your homeboy. I'm your professor.

With White colleagues, Black mathematics professors may feel a spotlight effect as well:

> [I was] working a summer program that I had worked for several years, a special math program for undergrads. And a few of us on the program faculty were just sitting around talking and I had had this question that was pertaining to mathematical induction, just a very nuanced sort of question. But I had had this question in my mind, I don't remember even what it was exactly now but I remember, I guess, something we'd be working on, preparing for the students, but I remember having this question. But I also remember feeling I could not ask this question. Because, you know, it's math induction, why would you [need to ask]. . . ? We were sitting around, it was the day the students were checking in actually, so there was a lot of dead time and two of my colleagues, you know, both White, one of them actually raised the very same question and they had this spirited discussion about it. And actually I still think it would have been different if I had asked the question, because they would have been like, instead of spirited debate about it, "Of course, it's such and such." And I remember standing there listening, of course eating it up because I had wanted to [know] too, but also frustrated that I felt I couldn't have asked this question.

Most of the mathematicians interviewed are, or have been, faculty in academe, with some noting that the transition between graduate school and faculty positions can be difficult. As one third-generation mathematician notes, "I think special education is required for a smooth transition, and in the end, assimilation in the culture and the practices of mathematics in an academic setting." One senior mathematician, who became a full professor after a long career in industry, notes "I always felt a little bit better protected than a young person just coming right out of graduate school. I wish that schools would assign brand-new faculty members, young faculty members, a mentor who can take them under their wing, talk to them about their research, write joint papers with them, eat lunch with them, just keep them from being isolated."

Black mathematicians in this study were asked to define their professional priorities, and most academicians said that research (including publishing and securing grants) was their most important priority, with teaching and service following closely behind depending on the type of institution where they were employed. Second-generation mathematicians noted that the emphasis on research, especially at small colleges, has become significantly more pronounced for young faculty than it had been when they first became professors. The areas of research in which Black mathematicians are represented are varied, with slightly more than half of the participants engaged in areas of applied mathematics, with the remaining in pure mathematics fields. Some participants felt very strongly about their choice of direction, with one noting:

> I was one of those students who was like, "I want my math to do something." I love working with numbers, but how can my numbers really help? That was the goal. That was the motivation for me becoming an applied mathematician. . . . I wanted to see my math do something in the world. . . . I felt that that was the key to getting more students of color interested in and motivated in and understanding and learning math better.

Others felt equally strongly about their work in pure mathematics and the necessity of showing young people that math was more than a "tool kit." One mathematician noted:

> You can always work on other people's questions. You're good at that. But what do you have inside yourself? It's the same feeling—I took up finally, I don't know why I waited so long, the piano. I was doing a little concert and there were actually people reacting favorably to the things I'd written. Because that was what I could do. Maybe it's not, you know, Ellington, but it's Me, whatever that is, and I get some pleasure from it. Other people apparently do, too.
>
> It's the same thing with mathematics. People say, "Mathematics is a young person's game," or whatever. It's something that's true, and I think that primarily it's because a young person has something original to offer. There's something in the way they're wired that

> allows them to see the universe a slightly different way than anyone else. They need to get that out. They need to ask their questions, because perhaps no one else is going to ask those questions. Perhaps they're the first person who can answer them.
>
> It's the same thing with music. After a while, people say, "That mathematician has nothing more to give." Well, he has more to give, but it's pretty much the same. His originality's been shown; his perspective now is clear. Maybe other people are starting to think that way, and he's tapped out. It's not because he was young, it's because he mined what he had to give. So I think I was able to mine what I had to give in mathematics, and I've seen the same thing happen in music.

The mathematicians who described their philosophies of teaching at some length all noted the importance of teaching students how to think. One professor suggested that better teaching occurs at smaller universities and colleges, because in large and/or elite ones teaching and learning were "more grade focused, less education focused, more entertainment, less knowledge transferred," and he lamented the focus on entertainment in teaching:

> I think there's a focus on entertainment in teaching, or an expectation, and that expectation is hard to meet in, you know, with a population of students [who are] innumerate, like don't have much proficiency with math. Like it's hard to make it fun, if you don't know the rudiments of it. If you have to tell a joke in English, and the person doesn't understand English, they're not going to think the joke is funny.

Faculty recognize that students have difficulty with mathematics, sometimes with fundamental and basic concepts:

> Some of the students here don't have the mathematical background that they should. So I had to understand that and make adjustments in my teaching, but I refuse to make adjustments in my requirements. . . . But I know that there could be a variety of reasons why they may not know [certain things]; especially because some of these people come from poor counties where their math teachers may have [had] a degree in physical education. So I have adjusted in that

> sense. But I know that it was because of [my undergraduate advisor's] strict demands [that we became successful] so I will not water down my curriculum. At the end, they are going to be competing with people from every institution, so they have to get the same. It is unfair to them for me to change the requirements, which around here, some professors do.

Stuart Potter, whose career has spanned jobs in industry, at predominantly White institutions, and, currently, at a historically Black college, describes this phenomenon as part of the larger mission of the institution:

> They have deliberately carved out that we are the urban university, which means that we said that this university presents opportunities to those students that wouldn't have any opportunity elsewhere, in particular students of color, African Americans. What that means, though, is that basically there are a lot of students whose backgrounds are incomplete.

He further notes that this is, in his opinion, the byproduct of an educational system in which students have been "passed along" for 12 years. When they arrive at his university, and faculty begin requiring more work, thinking, engagement, and effort, students may balk:

> So I tell my colleagues that [the students] have had 12 years of that, of how the game is played. Suddenly they get here and their view is that we are changing this game. We are changing a lot of the rules and they are pissed about it. "How dare you change the game? Don't tell me this. We know how this goes. Why are you changing it?"

In his work at this institution, Potter acknowledges that students may not fit the profile of college attendees, and some faculty can't get past that:

> There is one student here. . . . I remember when I first saw him. The terminology that comes to mind is "knuckle-osity." You know? He had the typical mannerisms of an urban youth, even the dress and all that. Watching him carrying on, I found out that he was transferring as a math major. I said, "Oh, God. This is a real knucklehead." He was in my foundations course; all the math majors have to take

> it. I asked him a question [in class] and found out from his response, "This guy can think." With that, I started going after him because I realized that he could think. In that course, I challenged him.
>
> Along the way he had felt he was getting some hostility from the other faculty members. I said, "Well, yes. You present yourself like a gangster. What do you expect?" But he noticed in my course [that] I didn't let him get away with stuff and he liked that. So towards that, he began to have an epiphany and he began to change. His whole attitude was different. He was dressing differently and talking differently and all of that. If I had reacted to him like everybody else had reacted to him, we would have lost him. I try to tell [my colleagues], "Yes. They don't have the best background." A lot of our colleagues think that we should take them all and ship them to community college and let them learn. You can't do this. You have to realize and you have to be open because you never know.

Some mathematicians have developed their own philosophies of teaching as they gained more experience in the college classroom, but for many, their own experiences as students have influenced how they think about the craft of instruction. A number of mathematicians were strongly influenced by high school teachers and their undergraduate professors. Similarly, as faculty, they feel a strong need to "give back" and encourage students to participate in mathematics.

Regardless of the type of institution by which they are employed, mathematicians' most often stated third priority and significant part of their work was ensuring that additional students—particularly Black students—became interested in mathematics. As one mathematician saw it:

> I think I'm an agent of change. I want to change how we as African Americans view mathematics. I think—I told my boss this once—as a group, African Americans as a group, wouldn't it be neat if we just took this up, just said, "This will be how we are identified"? Like when people look at, say, Asians or Indians, they look at them as the body that's providing the engineering capital of the future. Wouldn't it be neat if we were able to change things in such a way that they looked [at] the African American community and said, "They are going to provide the mathematical capital that this country needs for the future"?

Another third-generation faculty member noted that being a Black faculty member within a predominantly White faculty created particular responsibilities:

> At this institution, [there are only] two African American faculty members. And we're both new. But before that it was really sort of sad for the few Black students that they had not seen any sort of representation of who they could become. And I feel like having been there for a year I've done so much more service because people are like, "Oh my God. There's a Black woman on faculty [here]. You know? Or, "Can you mentor __?"

Some recognize that their commitment to students and recruiting might compete with their research agenda:

> Probably, definitely the first important thing for me is this continuing to add more African American PhDs and women to mathematics. Probably people will be saying that I am too young for this to be my most important thing, but because I was so influenced, I feel obligated to pass that on. So to get students excited about math and about deep math, that is my main focus. After that is probably the research. I know that if I can't keep it up, I can't pass it down because my advisor did not do research. [My advisor's] life is not quite fulfilled because of that. For me, I realized that that is something that I need to do.

Although many mathematicians felt this very strongly at the outset of their academic careers, noting the need to "pay it forward" or recruit others into mathematics early on, others came to this realization later:

> Actually when I first got to grad school, I wanted to just go into the industry and make money. When I was taking math courses and there was a course called operations research, the professor was like, "Yes. You can make pretty good money going into OR [operations research]." I was like, "Okay. That is fine. That is what I want to go into then." . . . One of the professors actually had a software company. He was a millionaire. He still is. I was like, "I want to be like this guy. I want to make good money and go to the industry."

> But going through the program [with my doctoral advisor]. . . . He was really supportive of minorities and minority issues and tried to support minorities getting PhDs and going into academia. That kind of influenced me to say, "Hey, I can make a difference in this world, too. I just can't really think about myself; [I should] try to help other minorities get PhDs and have a better experience in life." That really inspired me to be a professor.

Black mathematicians describe the issue of underrepresentation of Black mathematicians as a pipeline issue—many involved in recruitment and retention issues focus their service efforts on undergraduate and graduate students, and several do work with younger students. One second-generation mathematician with significant high school mathematics teaching experience described his long-time relationship with the public schools in the city in which he works:

> I really feel a connection to it and I continue to work with the schools. I basically feel I identify with the public school system because I went through the public school system and we are kind of active in that. . . . I am running now a program, actually at one of the high schools to get more minorities taking AP calculus. I've been doing this for the last six years. Before that I worked with the Algebra Project. I've worked with other efforts to sort of beef up the curriculum. So I've been, I'd say over the past 15 to 20 years, in sort of a continuous relationship with the city schools.

Other mathematicians have also developed or participated in outreach programs for elementary and secondary students. One "retired" mathematician speaks very proudly of a program he developed targeting African American students in the 1990s:

> I think the thing I'm most proud of is a lot of the work that I did in my summer programs for middle school and high school students. And I did that for about 12 years. And I get letters all the time from [former students]. . . . We started with them in the seventh grade and we were bringing in about 80 students every summer. And some of them could come back twice. But I'm getting letters all the time from kids who are PhDs. They're college teachers. They're doctors.

> They're lawyers. And they all say that our summer experience was one of the things that helped them a lot.

Some third-generation mathematicians have been doing this work with younger students since their college or graduate school days. One mathematician discovered a prior connection with one of his college students:

> This past spring I was teaching real analysis and I noticed one of my students had Oakland as part of his email address. So after class one day, I said, "Are you from Oakland?" He said, "Yeah," and so I said, "Do you know Washington Elementary?" and he perked up, "How do you know Washington Elementary?" And he asked how did I know, I said, "I used to do some work there while I was in graduate school."
>
> He paused, and he said, "Did you have a chess club?" And I did. And my chess club was so much fun, this was during the time I guess I wasn't officially there anymore but I'd still give a little time and one of the things I'd do, I started this before-school chess club. And it was cute because the teachers would walk by and they would just smile because here are all these little kids, mostly boys, it was a primarily Black school. I had a couple of girls who came in too but it was something about the boys and they came. And they were just well-behaved and everything, and you'd see the teachers come by and they'd just look in and smile at us in the morning and then go on to their classroom. But yeah, we had fun. We had a little tournament at the end and I save everything, which is why my office and my life are a mess, because I keep everything and can rarely find it. Anyway, I found the old tournament scorecard and there was his name. And so I took it to class one day, showing it off. And so he'd been in the club both years. And he still plays chess, and he's [here] now. Math major.

Several mathematicians interviewed served in administrative capacities at their institution—as department chairs, deans, vice presidents, and/or directors of special programs. In their roles as administrators, Black mathematicians wield power to effect practical and policy changes that have the impact of increasing Black student representation at their institutions and improving retention of graduate students, when they are at primarily White institutions

(PWIs) as well as at HBCUs. As a former mathematics chairperson at a large PWI said:

> Even though they were here and we had brought Black students into the department they really didn't know each other because they had come [to see him] independently. They didn't talk to each other, come in a group or anything. They just each came in individually. . . . That is when I realized that this is a predominately White campus. Everything in sight was White. I think a lot of the students were, sort of epiphany, they were trying to join the White networks as opposed to also interacting with other Black students. They started doing that, and we started doing that regularly. Every year we would have at least one get-together. Then when I was chairman I actually thought of bringing them up into my office so we could talk about whatever was going on. That way I figured they got a chance to institute some changes to the graduate program. They were concerned about how that was going to affect them. Well, they could ask me directly. They didn't have to go ask the graduate chair. They could find out from the horse's mouth. At least what we were thinking. They still might not have believed it, they might not agree with it, but at least they got to hear what I was thinking in terms of why we were doing this. I also brought in some people who were Black mathematicians at other places so they could see what kind of jobs they could get—to do some networking with other Black professionals. I think it just helped them to see that they could interact as a group.

When Black faculty are not department chairs per se but high-level administrators, they find themselves called upon in different ways by their institutions in ways that have nothing to do with mathematics:

> The other parts of the job [are] because I am a Black administrator on campus. That means that all major Black problems are mine. And of course along with that, if they got to this level they are not easy to solve and that they are big headaches. The incoming Black faculty is one of my areas of passion. And I like to try and keep our faculty diverse as much as possible. Staying with the times. Stay on top of things as department chairs. I like my presence when I come into a room to make them think about whether or not they have done the right thing to be serious. I have my own approach to it and I don't

like to be confrontational by nature, only when I think it helps the cause.

I still teach a class, and enjoy that. And administratively, that is smart of me because when a new president comes along and cleans house, you have to go back to your department. You want them to know you. So I maintain contact with the department. The pure research part has suffered tremendously. I don't do much of that anymore, but the learning aspect of it I still like. I keep abreast of my colleagues to keep up with the field.

Conclusion

Black mathematicians of all generations feel keenly their underrepresentation in mathematics. They have varied approaches to contributing to the resolution of this issue—some as professors target undergraduate students and mentor them, encouraging them to pursue a doctorate. Others, as administrators, actively recruit Black students to graduate programs and seek to develop structures and practices within institutions to retain them. Still others develop national programs as well as programs housed within individual institutions that are designed to recruit and retain talented students in mathematics. Countless others describe informal and formal mentoring relationships from which they have benefited and which they engage in with students themselves. Black mathematicians describe a variety of mechanisms and relationships through which this occurs—they describe warm and supportive relationships with multiracial faculty within institutions as well as programs and initiatives spearheaded by Black mathematicians' organizations. Through these relationships, they develop formal and informal networks that are intergenerational and extend to college students as well as to secondary school students. Mathematicians report a variety of working relationships with fellow students in graduate school, and for many, the kinships and communities that they develop during these years are integral to their success.

Black mathematicians who have succeeded in earning their PhDs and who have joined the profession as faculty members have described various instances and situations in which their race (and for women, their gender) has been an issue. In addition to how Black mathematicians are viewed by others, a kind of spotlight effect, on occasion, has been noted by Black mathematicians, in which their behaviors are affected by their profound awareness of their Blackness in particular contexts. Although significant progress has been made in terms of their inclusion in the profession, they continue to encounter discourse and practices that attempt to exclude them from the mathematics enterprise.

Underlying some narratives is the theme of sacrifice. Mathematicians tout research as being the most important aspect of their work, but a few describe their commitment to developing more mathematics students who are from underrepresented groups as so paramount that their research activities may suffer by comparison. They report that they understand that their research (and for junior faculty, becoming tenured and getting to "hang around," i.e., remain at their institutions) should be the most important aspect of their work and that there are multiple avenues to getting young people involved in mathematics. Some are more active in these activities than others. This tension is one that has been felt by some Black mathematicians throughout their existence in the United States.

Recognizing that the representation of Blacks in mathematics is a pipeline issue, several mathematicians describe efforts by Black student organizations and historically Black colleges and universities to engage with elementary and secondary students in communities and in schools. Further, professors at these institutions are invested in developing Black students' mathematics interest and talent in college and supporting them as they pursue graduate study. As one second-generation mathematician noted, the chairperson of the mathematics and science department at his HBCU, very proud of his own PhD, was both proud of and yet rueful about the fact that he was one of "only 50" Blacks with a doctoral degree in his field at the time. In the next chapter, I explore how historically Black colleges and universities are a critical resource for the recruitment, retention, and success of Black mathematicians.

FIVE

FLYING HOME

Historically Black Colleges and Universities and the Development of Mathematicians

Although some attention has been paid to the success of historically Black colleges and universities in facilitating access to science careers for their graduates, the mechanisms by which Black colleges promote this success have largely gone unexplored in the literature, with a few notable exceptions (Hilliard, 1995, 2003; Scriven, 2006; Southern Education Foundation, 2005; Tucker, 1996). These mechanisms are also often viewed as static, without emphasis on the traditions and practices that influence succeeding generations of mathematics students and faculty. But the narratives of Wayne Leverett, Eleanor Gladwell, Laverne Richardson, Stanley Parker, and others clearly demonstrate the impact that Black colleges, their faculty, and extended networks of faculty, students, and alumni have on mathematicians' professional lives. For some, these experiences are so seminal and impactful that mathematicians who attended HBCUs feel a strong urge to "come home" to be professors at their alma maters or other HBCUs to "help continue that tradition" of mathematical excellence.

In this chapter, I use the notion of "mathematical spaces" (Walker, 2012b) to frame Black colleges' success in attracting and contributing to the retention of Blacks in the scientific pipeline (specifically in mathematics) long after graduation. In particular, there are characteristics of Black colleges that mark them as spaces (Lefebvre, 1974; Soja, 1989) rooted in social and cultural geographies within specific historical contexts that support mathematics development, incubate mathematical talent, and disseminate effective practices beyond the walls of these institutions and across generations of students. Thus, I have defined *mathematical spaces* as sites where mathematics knowledge is developed, where relationships or interactions contribute to the development of a mathematics identity, and where induction into a particular community of mathematics doers occurs (Walker, 2012a). These spaces, then, may be physical locations like a school or classroom, and/or they may be locations to which the individual attaches a particular social, cultural, or mathematical meaning due to interactions and experiences she or he had there. These spaces are not specific or unique to Black mathematicians' development. However, I argue that there are unique cultural, social, and historical parameters (O'Connor, Lewis, & Mueller, 2008) and practices (Gutierrez & Rogoff, 2003) that influence how Black mathematicians and others engage within and construct and maintain Black colleges as mathematical spaces. In the sections to come in this chapter, I describe the work of Black colleges as recounted by Black mathematicians in promoting mathematics excellence in general, focusing mostly on the practices of faculty within these institutions. Then, I describe how two institutions in particular—Morgan State College (now Morgan State University) and Spelman College—exemplify spaces as sites for mathematics knowledge and identity development, socialization, and induction. I suggest that Black colleges, their faculties, and the mathematics practices that arise within the college environments have had lasting effects not just on their graduates but on the field of mathematics in general. At a time when some question their relevance in a supposedly postracial era, there are important lessons to be learned from these institutions, which have traditionally promoted education for the underserved as well as fostered academic excellence and increased access to professions for African Americans.

Researchers have pointed out that "traditionally Black institutions produce disproportionate numbers of students who persist towards doctorates in the sciences and engineering, even though most African American students attend predominantly White institutions" (Cooper, 2004, p. 539). Indeed, the "eight colleges and universities that produced the most African-American graduates who went on to earn PhDs in mathematics or science in the 1997 to 2006 period are all historically black colleges and universities" ("Black colleges and universities," 2008), with Harvard University and the University of Maryland ranked ninth and 10th, respectively.

Howard University, the top-ranked producer of Black PhD graduates, has a long tradition of excellence in mathematics and science, with its faculty at one time including several of the first Black PhDs in mathematics (including

Elbert Frank Cox, the first Black person to receive a PhD in mathematics, as well as David Blackwell, who was interviewed for this book). Bridglall and Gordon (2004) call Howard, in the years 1927–1960 under the leadership of President Mordecai Johnson, "an oasis for African American scientists" (p. 338). A chairperson of Howard's mathematics department, Dudley Weldon Woodard (1881–1965), was only the second Black American to receive his doctorate in mathematics, from the University of Pennsylvania, in 1928. After completing his undergraduate education at Wilberforce in 1903, he received another bachelor's and a master's degree from the University of Chicago. At the age of 47, after completing "advanced mathematics courses in summer courses at Columbia University," he completed his PhD from the University of Pennsylvania. Woodard's career was restricted to historically Black colleges and universities; after completing his PhD, he returned to Howard, where he had been a respected administrator and teacher of mathematics, and started an MS program in mathematics there in 1929.

Howard was the first HBCU with a doctoral program in mathematics (beginning in 1975) and is still one of only three HBCUs with any doctoral program in mathematics.[1] With Black colleges continuing to provide a significant share of the Black science and mathematics majors who enroll in graduate schools, it is important to consider the genesis of these institutions and the implications of their existence.

Rooted in a strong community context for higher education, many Black colleges began as fundamental education programs designed to address the lack of literacy and formal education of Black Americans emerging from slavery and a Jim Crow existence in the segregated South (Anderson, 1988; Drewry & Doermann, 2001). Hampered by lack of funding support and philosophical disagreements with state governments and foundation revenue sources about the appropriateness of a classical university education for Black Americans (Donaldson, 1989), Black colleges often struggled to develop research programs but demonstrated an unprecedented commitment to teaching and developing a professional class of Black Americans (Pearson & Pearson, 1985; Winston, 1971). Often founded in collaboration with churches, missionaries, philanthropists, and committed community members, the lingering mottoes of Black colleges reflect their spiritual beginnings and their continuing commitment to community service and improving life chances for their graduates and the Black community on the whole—"If the Son shall make you free, ye shall be free indeed" (Lincoln University), "truth and service" (Howard University, North Carolina Central University), "Our Whole School for Christ" (Spelman), "And then there was light"

(Morehouse). It is thus not surprising that many HBCU faculty describe their work at these institutions as a "calling," as though they were "called" to teach there in the same way that ministers are called to preach the gospel.

These institutions clearly did not arise in a vacuum. It would be ahistorical to consider them, largely located in the southern United States, without acknowledging the importance of their being situated within a unique milieu that is "historically considered the reservoir of African American culture in the nation" (Morris & Monroe, 2009, p. 21). That Black colleges existed at all in the South—providing a paradoxical example of Black intelligence and success when the region was mired in Jim Crow mores and practices and legal discrimination that underscored a White supremacist ideology that Blacks were inferior in every way and most certainly intellectually—cannot be underestimated (Winston, 1971). Black colleges existed in many ways as imperfect islands of opportunity by necessity, not choice, for Black Americans across the South when Black opportunity was circumscribed legally and politically.

The theme of Black colleges as sites for mathematics learning, professional mentoring, and the development of generations of mathematicians emerged repeatedly in mathematicians' oral narratives—including those of mathematicians who did not attend Black colleges themselves. Their accounts and the historical record of Black colleges' success in producing graduates who excel in the sciences stand in juxtaposition to their remembrances of those who questioned the quality of Black colleges and to the ongoing debates about whether or not Black colleges are needed in the 21st century.

Eighteen of the 35 mathematicians interviewed attended historically Black colleges or universities (HBCUs), with two of those initially attending predominantly White institutions and then transferring to historically Black colleges. The mathematicians who attended HBCUs in this study are fairly equally distributed across generations. Blacks, in whatever era, have always attended predominantly White institutions for their baccalaureate degrees. As described earlier, the first Black male and female PhD mathematicians in the United States, Elbert F. Cox (1895–1969) and Euphemia Lofton Haynes (1890–1980), attended Indiana University and Smith College for their undergraduate degrees—both predominantly White institutions. The three mathematicians interviewed for this study who earned their PhDs in the 1940s and 1950s—David Blackwell, Evelyn Granville, and Clarence Stephens—attended the University of Illinois, Smith College, and Johnson C. Smith College (now University), respectively, for their undergraduate degrees. Thus, Clarence Stephens is the only mathematician of the first generation interviewed for this book who attended an HBCU.[2] After Blackwell, Granville, and Stephens had earned their PhDs from Illinois, Yale, and Michigan, respectively, all three began their university careers at Black colleges, due to the fact that White colleges were extremely unlikely—and largely unwilling—to hire Black faculty members (Donaldson & Fleming, 2000). Winston (1971) notes that until the last quarter of the 20th century, there was a "virtually impermeable racial barrier [that] excluded Negroes from white universities and their superior facilities for teaching and research" (p. 678). Although a number of Black mathematicians of this era eventually were able to work at White institutions, the role of Black colleges in facilitating networks and professional opportunities for succeeding generations of mathematicians remained, and continues to remain, critical, as this chapter will demonstrate.

After teaching for a year at Southern University in Louisiana and Clark College in Atlanta, David Blackwell's 10-year career at Howard began in 1944. As he told an interviewer, "it was the ambition of every Black scholar in those days to get a job at Howard University. That was the best job you could hope for" (DeGroot, 1989, p. 592). He described his teaching load as substantial and including both undergraduate and graduate students. By his third year at Howard, he was the department chair. When interviewed, Blackwell did not recall any of his students from Howard going on to earn their PhDs in mathematics, "although I had some very good students." He was later recruited by the University of California, Berkeley, and retired from there in 1988.

Evelyn Granville, after completing a one-year fellowship in New York City at New York University's Institute of Mathematical Sciences, began teaching at Fisk University in Nashville in 1950. She taught there for two years,[3] noting:

> And of course we had very well-prepared students. In fact, two of the students, actually three of the students I taught—now I don't take responsibility for it, but just to show you the caliber, three of them went on, two got PhDs in math,[4] and one got her degree, her doctorate in education. But I can't take responsibility because I was only there two years, but that indicates the kind of academic training that students had when they came to Fisk.

After teaching at Fisk, Dr. Granville worked for the government and private industry, eventually teaching as a faculty member again in California at California State University, Los Angeles, and other institutions.

Clarence Stephens began his university teaching career at Prairie View University in Texas, but soon left because it was in a rural area and he needed access to a university library so that he could finish his dissertation. He ended up at Morgan State College in Baltimore and was promptly made the department chairman—at 30 years old. He spent 15 years at Morgan State, leaving in 1962 to become a faculty member at SUNY Geneseo. Dr. Stephens's work at Morgan will be described later in this chapter by his former students and in his own words. Dr. Stephens reported that his leaving Morgan State was largely due to two reasons: his wife had grown up in the countryside and disliked urban Baltimore, and also

> There was a chance to get into an integrated situation. You know, where we'd been kept out of—and I decided to try it out.

Dr. Stephens was immensely successful at Geneseo and, later, Potsdam, and he remained a faculty member in the SUNY system in upstate New York for the rest of his academic career.

Black Mathematicians and HBCUs

The discourse in the Black mathematicians' community about Black colleges largely characterizes them as focused on teaching and service to the Black community; as important social, cultural, and historical spaces where one's intellectual ability is not subjected to mythologies about Black inferiority; where financial and research resources may be limited; and as institutions that are consistently undervalued and denigrated by some as "not as advanced" or "lower in quality." Numerous studies of HBCUs describe them as institutions that are often lacking in financial resources but committed to the mission of educating African Americans, preparing them for professions, and contributing to the uplift of African American communities. Some have prided themselves on attracting excellent students as well as successfully matriculating those students whose academic records precluded them from attending other institutions and proving that these students can excel at high levels when provided with quality education and mentoring (Drewry & Doermann, 2001). The Black mathematicians in this study who attended HBCUs attended private institutions such as Morehouse, Howard, Xavier, and Fisk, as well as public institutions like Alabama A&M, North Carolina

A&T, and Morgan State. Regardless of the institution type, mathematicians described common themes relating to their mathematics experiences within these institutions. These themes, broadly, center around the overall ethos and mission of Black colleges, which support the development of well-prepared graduates, strong family and community legacies and ties, and high-quality teaching and mentoring that is both rigorous and nurturing. These themes are found in literature about effective practices of HBCUs in general, but for mathematicians they have special relevance to their study and practice of mathematics.

Ethos and Mission

Black mathematicians often describe the overall milieu of Black colleges as important to the development of themselves and other students, regardless of field of study, as intellectuals. Many spoke about the commitment to community uplift exemplified by Black colleges, and they themselves felt compelled to live up to that mission. For example, as Elizabeth Ricks, a third-generation mathematician, recalls:

> I remember being president of the math club. I was really big into community service. I organized this program called In School Tutoring. We had an elementary school that was in the neighborhood of [the college]. Of course, the neighborhoods around Black colleges tend to be the poor, not so good neighborhoods. So I organized [our college] students to go over there in the middle of the day to help students while they were in class. We would either help them with reading or help them with math. The teachers were really receptive to our coming over.

In addition, as the mathematicians in this book relate, the high expectations of the faculty, staff, and administrators for students' success are important factors in promoting persistence in the study of mathematics. Reflecting on his experiences at his alma mater, Stanley Parker suggests that although he probably would have done well wherever he went to college (he was accepted at Ivy League institutions and prestigious engineering schools, as well as by his HBCU alma mater), he would not have received the same "push" to pursue graduate study at White institutions:

> [At my HBCU alma mater] the expectations were "you're so good, I assume you're going on to get a PhD, aren't you?" . . . That sort of expectation, as opposed to just "you should be fine in the job market"—you're lucky if someone takes that extra interest in you to see the potential and pushes you to obtain it. I think that's what I especially hold dear and what I still think we do for our students that just doesn't happen [everywhere]—I mean, every now and then, but not with any regularity like we do here.

Similarly, when asked by a major professor at his predominantly White graduate school if he regretted going to an HBCU for his undergraduate education instead of a "place like [this one] so you would be much more advanced," a second-generation mathematician replied:

> If I had gone to this place [his graduate PWI] for my undergraduate education I wouldn't be here today as a graduate student. . . . I had to get anchored in what I was doing where knowledge was the issue and I didn't have to deal with the politics to get through undergraduate school.

Stanley Parker challenges the notion that HBCUs provide a weaker academic environment than majority institutions. Having participated in a popular 3–2 professional program between an HBCU and a predominantly White institution, he recalled:

> When we got to [the predominantly White neighboring institution] we realized we had gotten more calculus with Dr. Ricks [at his HBCU] than they got there. He just demanded and expected more and did certain things they didn't do. They talked a good tough game, they flunked a lot of students out, but that said: there are times when you have a responsibility to the entire class, that you can't just race ahead and leave everyone else in the dust.

Many HBCU graduates feel very strongly that HBCUs provide an academic, social, and cultural experience for Black students that is unobtainable anywhere else. In addition to administrators, faculty, and staff contributing to these environments, a number of HBCU graduates describe strong communities of students at these institutions, particularly in the sciences. Sometimes these communities were built deliberately: As Russell Means previously stated, an administrator, Ralph Lee, recruited several top students from Means's high school in Montgomery to attend Alabama A&M. Clarence Stephens did something similar at Morgan State, recruiting from Douglass and Dunbar High Schools in Baltimore. In more contemporary times, Florida A&M and other universities (including the Meyerhoff Scholars Program at University of Maryland, Baltimore County (UMBC), designed and implemented under the leadership of Freeman Hrabowski, himself an HBCU graduate) have also recruited cadres of strong Black students in the sciences. In other scenarios, groups of students create strong peer communities themselves, with the support of faculty, administrators, and staff. As one mathematician recalls:

> I would get the class together and we would meet. One of the students was older so she had an apartment. We would meet over there and have pizza and do math. The chair would leave the building open on like a Saturday so that we could do math problems and all of our upper-level classes. We were very close as far as a class unit. One of the students was from [a neighboring town], so we would go to his house and study. His parents would cook dinner for us. . . . Because the classes were small, it was kind of like, "We are all going to get through this." That is one of the things that I definitely like about my HBCU experience. . . . There was probably a core of at least six or seven of us that all took classes together. After that proof class, we would move together. We definitely intentionally made our schedules alike and decided, "Okay. Are we going to take this class this semester or are we going to wait until next semester?" . . . Because my year was so successful—usually they had this one-hour seminar class where they kind of just talk about what you have learned, but because they thought that we were kind of advanced, they actually taught a course in topology, which had not been ever offered at [the institution].

As will be recounted later in this chapter, graduates of Morgan State and Spelman College also describe in great detail enriching, competitive in some cases, and supportive student communities.

The characteristics and combinations of rigor and care, service and community, and the opportunity for a unique cultural experience are a draw for a number of mathematicians, particularly those who attended predominantly White high schools.

> Looking back, I only applied to historically Black colleges. I guess I wanted the experience. My parents didn't go to historically Black colleges. I went on a historically Black college tour, though, when I was in eleventh grade, of universities across the eastern seaboard. So I imagine that that definitely influenced me. One of my friend's mothers had gone to [the undergraduate institution Richardson eventually attended], so we had talked about that. I remember talking about that with her. (Laverne Richardson, interview)

For several mathematicians in this study—men and women, younger and older mathematicians—family histories include stories about the traditions at HBCUs as experienced by parents, uncles and aunts, grandparents, and other family members. Although it is certainly not unusual for college students to attend the same colleges or universities as their parents or other family members, for mathematicians, these institutions enter a particular realm of lore. For one mathematician, her father's attending an HBCU did not become particularly salient until they visited his alma mater on a family trip when she was a graduate student and she realized that he had been taught by her "hero," Marjorie Lee Browne:

> We were able to walk in, on a Saturday afternoon. We were actually able to walk into this building, like, maybe go up a level, walk down this dark hallway. And Dad is saying, "Oh yeah, this is where I took math classes." The math department was still there. They had a glass case of this beautiful portrait of this Black woman in there. I look at the picture and I know immediately that it is Marjorie Lee Browne. Just because from going to Spelman, I don't know, you just knew these names. I'm, like, "Oh my God. Oh my God. Why is she here? Oh my God." And I am thinking I am telling my parents something that they don't know. But my dad's like, "Oh yeah. When I was here I had her. I was a math major." I had not known this.

High-Quality Mathematics Teaching and Mentoring

Mathematicians across the three generations speak of the high-quality teaching and learning of mathematics that has occurred and continues to occur in HBCUs, as supported by the literature (e.g., Southern Education Foundation, 2005). The commitment to preparing excellent graduates in mathematics (and other scientific fields) begins early—several HBCU graduates described intensive pre-college or undergraduate summer programs designed for mathematics and science majors, including those at Spelman College, Howard University, and other institutions. As one mathematician who attended Xavier University in Louisiana said,

> I got into a summer program called SOAR [II], for Stress on Analytical Reasoning. SOAR I was for the biology and chemistry people and SOAR II was for the engineering, math, and computer science people. It was a six-week summer program for all of those who were admitted to Xavier. I think we took the SAT at the beginning, then we were doing practice SAT and then we took it at the end. But we had quiz bowls and we had to learn vocabulary. It was a very intense program, but it was so much fun. So that's what sealed it for me. I knew I had to go to Xavier because of that program. . . . The goal was also to help you transition to college. We took Pascal, so we took a programming class. I think there might have been a math class. I think we took two classes, but I can't remember the other one.

Once Black mathematicians enrolled in college, they experienced (and recalled vividly) rigorous mathematics environments. Attending a public HBCU in the South, Russell Means describes the mathematics faculty:

> We were terribly impressed. We got pretty good instruction. There were maybe only two persons with PhDs in mathematics at the time at [our college]. One of those two was a Cuban. He was our best teacher. He taught us quite a few courses and he was really excellent. . . . But he couldn't teach us everything, so some of the other courses were taught by people with master's degrees. Everybody had to have at least a master's degree. A couple of teachers went back to school and now have their PhD, but they didn't have it at the time that I came through. Another one who went back to school was a dynamite teacher. He subsequently became chair of the math department. When we were there, he had a master's degree. He taught me differential equations. . . . Most of the teachers had master's degrees in math, with the exception of those two people. They were pretty good teachers. So I don't have any regrets about that.

Speaking specifically about her classes, a second-generation mathematician who was a Spelman graduate focuses on their challenging nature:

> [Classes] were challenging, very challenging. The usual, calculus and pre-calculus, even though it wasn't called by those names. And the standard curriculum, I was starting with pre-calculus, calculus, topology, linear algebra, abstract algebra, and I guess number theory, I'm not quite sure, I can't remember that far back.
>
> Q: What were the professors like at that time at Spelman?
>
> A: They were very good, you know, most of the professors required proofs from year one, as freshmen, so, you know, the courses were very challenging for all students because I don't think any of the students came there ready to do proofs.

Finally, a mathematician from the third generation describes her experiences at her HBCU alma mater:

> The calculus I, II, and III courses were very large because of the engineering [emphasis]. At that point, you were just a number or whatever. Once you got into any courses that were just for the majors, they were very small and very nurturing for me, but a lot of work. I credit my advisor realizing what you had to do in order to be successful. We worked a lot. The transition course, to me in my opinion, is kind of "make or break" as far as whether or not you are going to be successful in graduate school and get that good foundation. That course was not there before [my advisor] came. She got them to implement that course.

Other HBCU graduates describe the rigorous environment, and the willingness of professors to go above and beyond the regular curriculum:

> [My chemistry professor] would have problem sessions on Friday night from 7:00 to 9:00. And only about 10% of the class would come. But you also saw that those were the same 10% who were setting the curve because not only had he taught well in the classroom, but he answered any question we had and showed us any mistakes we were making in our assignments, how we should be doing it. That was chemistry. In mathematics I had a professor who was so good at teaching that he became a legend in his own right. And there were

> those who were saying that if a wayfaring fool wandered into his class and sat down and paid attention that he would learn mathematics. He made mathematics come alive. And he taught me, he said, "if you really understand mathematics you can sit down in a restaurant and explain it to somebody on a napkin, the concept that you know about."

For many mathematicians who attended HBCUs, the outstanding teaching and mentoring they received are inseparable. In the preceding examples focusing on teaching, there are key elements of mentoring—teaching students mathematics content and concepts, exposing students to habits of mind and study, modeling effective pedagogy, and providing opportunities to engage in mathematical research. Beyond the classroom, Black mathematicians describe key elements of communities at HBCUs that promote not just their mathematics learning but also their induction into the mathematics profession. Mathematicians in this study describe both formal and informal mentoring practices of faculty and staff at HBCUs that promote their development. These mentoring practices manifested themselves in significant ways, small and large. Conversations about graduate school—and which graduate school to attend—were reported by mathematicians often, with some mathematicians' major advisors encouraging their students to attend their own former graduate schools, as a path there had already been hewn. This was the case for Clarence Stephens as well as for several other mathematicians. Other undergraduate advisors shared their own experiences with their students as cautionary tales and made strong recommendations about how to choose a graduate program, where not to go, and with whom not to work. In addition, several Black mathematicians were invited by their HBCU alma maters to become instructors during summers or breaks in their graduate school careers. One student noted that her experience returning to her HBCU, which came after she had left her first graduate institution, was critical in terms of her reentering a doctoral program:

> By the grace of God or something, I don't even know what the initial contact was, I contacted folks [at my alma mater]. They knew that this was happening and they rescued me. They said, "Come on down here, teach class." So I was teaching and basically it was the day I came back, it was like, "You're going back to school, you're going back to school," and I was just sort of nurturing myself.

> The first year I'm like, "I am not applying," because you know, I showed up there that fall and I couldn't even think about filling out applications. I felt like such a failure. But they're like, "You're going back, you're going back, you're going back." So by the next fall, I applied and stuff like that. So I spent two years teaching [there], and that was the best thing that ever happened to me, because it got my confidence back up and, you know, teaching always helps with learning.
>
> It really sort of gelled, and they were just so supportive, they were just so, so supportive from that. And it reminded me that I could do it, all that craziness was not about you, there's no way I would have gone back on my own. Having that kind of support was huge.

In this way, Black colleges continue to keep ties to their graduates, but they also provide needed professional development and support and continued mentoring. Below, Russell Means further describes the key role of his major mentor at his undergraduate HBCU alma mater. The chair was not a mathematics professor but was nonetheless influential in Means's decision to pursue graduate study in mathematics:

> When we got to be seniors, then we got to interact with the chair a little more. . . . That's when he started preaching graduate school to us. It didn't matter that we weren't going in physics, we were going in math, as long as you go to graduate school. He almost insisted that we go. You know, you have a high regard for these guys. I said, "Okay. He wants me to go and get a PhD." So he's going to help you get into graduate school. He showed us how to apply to this school and that school. . . . He gave us the assistance in how to take the Graduate Record Examination. He mapped everything out for us. He would write recommendation letters to the schools to get you in. He pretty much took us under his wing. That's how we ended up deciding to go to graduate school.

Other mathematicians' mentors were their mathematics class instructors:

> [Taking that class] is when I met my advisor, Louisa Richards [another Black mathematician]. I think that she definitely saw something in me. She began to tell me that I really had a knack for

> mathematics and started telling me about summer programs that I could attend. . . . You know some people felt like she was very, very demanding. So in like her calculus I and II classes, she has the lowest enrollment. You know the rumors had already gotten out, so people did not want to enroll in her class. So she was very demanding, but I knew that because I like math, I knew that she was really challenging us. I just felt like she knew what it would take. She was genuine. We talked about other things, not just mathematics.

Edgar Tremain left his first undergraduate institution, a large predominantly White institution, in his junior year. On the recommendation of his uncles, who had attended an HBCU, he transferred that year to their alma mater:

> The other positive thing that happened to me while I was there was one of the faculty members, because everybody told me that you couldn't get an A in this guy's class. He was really that hard. As a matter of fact, the whole time he had been there he had never given an A in his upper-level classes. And I took [one of those classes] and I got an A. Then he told me personally: "Don't let anyone ever tell you that you can't get an A. You can always. You put your mind to it, you can always do it, you know, do what you need to do." Then the other thing he decided to do was to have me work on a project. So this is my first experience with working on a project. It was kind of a neat little thing that he was thinking about doing. [It] turned out that this problem . . . was a problem that these big mathematicians were working on. But I think really, to tell you the truth, in two space the answer was known. I think probably he told me that it wasn't quite known at that time—it was true for any Euclidean space, but my assumption at the end is it probably is because usually those things carry over pretty easily. But in any event, when I got to graduate school and I went back and looked at that problem I knew it was a problem of functional analysis. So it was kind of interesting that he got me involved in it. And then he got me traveling around giving talks on the results that we had. . . . He would call [several schools in the area] because he had friends. He was from Louisiana but he graduated from University of Arkansas. And so since he graduated from University of Arkansas he knew all these people that were teaching in these small schools in Arkansas. So he was taking me around to

> those schools, me and two other guys. We were giving talks. And then he also, they were also paying for us to go to conferences to give talks. So we were actually going to things like NAM conferences and giving talks and that sort of thing. People were surprised at these results that we had and then the last time I gave a talk I remember [a university] invited him to come out there to give a colloquium talk. And he said, well, what he would rather do is bring two of his students out there to talk. And so they said okay. And so he brought me and this other guy out there to give talks.

A number of Black mathematicians interviewed are themselves currently faculty at HBCUs. When asked to describe their priorities in their work, a key one is the notion of "giving back" and inspiring others to become PhD mathematicians or at least to pursue mathematics-related careers. Some have returned to their alma maters to teach as faculty members, and they describe the significant pressure they feel from faculty, administrators, and staff to rejoin these institutions.

> So I was visiting around the time I was finishing up my PhD, and I was getting pressure from the people that were here. The chair at the time—by the way he just retired from the university last year—kind of pressured me, he says, you know, "You got to come back." But the person who pressured me the hardest, which was really funny, was the department administrative assistant. She had been here for, like, 30 years. She kept saying, she would give me this whole thing about, "We need you here. You know, you're African American. Kids need to see you, they need to talk to you," and was giving me this whole spiel. I said, you know, "Okay. I'll come back at some point. But right now I am going to graduate. I am going to take advantage of a lot of opportunities. I am going to do a lot of things." . . .
>
> Somehow I just kind of gave in. I had offers at White schools that were paying more money and would get me better opportunities on some things and I knew that they would. But I said, "Before I die, I'll come back"; that was my plan. I don't know how I had, like, a moment of weakness and I said, "Okay." So I came over, taking less money, and I was angry about that. Really I'm still angry about it. It took a couple of years [to get over that], I'm still here. (Herb Carter)

Several Black mathematicians who are faculty at HBCUs describe their work partly in sacrificial terms—there are financial sacrifices made, often, when taking a faculty position at an HBCU. The teaching load is often higher at these institutions than at their predominantly White counterparts. But they indicate that the reward they receive from working with students and being part of these institutions that hold such significance in the Black community is substantial. Describing his work, one senior mathematician noted that his priority as a faculty member at HBCUs was

> to demonstrate by example and by demand to show these students at HBCUs that you could perform as well or better than any students in the world if you applied yourself. And I was offered a number of opportunities to teach at majority institutions that I rejected politely.

Younger faculty who wanted to begin their university careers at HBCUs reported being advised against it:

> For me when I graduated from [my graduate institution], I wanted to go back to [my HBCU alma mater] and teach. You know? I felt like they helped me so much and I wanted to go back and do for the students what [my advisor] did for me. I loved [the city], so that is what I wanted to do. But most of my mentors and my advisor, who was one of my mentors too, I love them to death. He is a White male. They come in all races, of course. All of my mentors told me, "Don't do it. Don't go back to an HBCU. They don't have as much money. You have to teach too many courses. Your research will fall." That type of thing. "You can do it later."
>
> So I got an offer from [my alma mater]. When the chair, who was my chair when I was a student, offered it to me, I actually told him, "Don't tell me the salary." I didn't even want to know because I felt like it doesn't even matter, like, "I don't think I should go back." So I went to a majority institution. It was 10% minority and that is all minorities, so 90% White. The students actually were great. But I did feel that I wasn't sure that this is what I was really meant to do. It was kind of like a calling type. Even though I loved the students and believed that they loved me and I was still the same person, I would have a class with no minorities; no Asian, no African Americans, none.

Eventually, after a two-year stint at a majority institution, Laverne Richardson accepted a position at a historically Black college. Another mathematician, Stuart Potter, always had a strong sense that he would end up at his alma mater:

> Even as a student it wasn't a secret that there was an eye on me to sort of come back. I certainly didn't have to and if I did there was still a question of when would that be, but there was sort of an eye. . . . Now I'm back.

Sites for Mathematics Knowledge and Identity Development, Socialization, and Induction: Morgan State University and Spelman College

Three mathematicians interviewed for this study—Earl Barnes, Arthur Grainger, and Scott Williams—attended Morgan State College as undergraduate students majoring in mathematics during the 1960s. They are uniquely positioned to describe the rigorous undergraduate mathematics program they and others encountered. Their experiences are inseparable from the influence of Professor Clarence F. Stephens, who was the ninth African American to receive a PhD in mathematics (University of Michigan, 1943) and whose own educational experiences in the segregated South—attending an all-Black boarding school, the Harbison Institute, and Johnson C. Smith College in the 1920s and 1930s—appear to have influenced his philosophy of teaching and learning mathematics. Dr. Stephens, who was a professor of mathematics at Morgan from 1947 to 1962 and later had an equally influential career at SUNY Geneseo and Potsdam, was also interviewed for this study.

Dr. Stephens's work at Potsdam has been chronicled (Datta, 1993; Megginson, 2003), but his work at Morgan State, where he first tried out these ideas, has been less fully described. In my interview with Dr. Stephens, he described his recruitment strategy for talented high school students in Baltimore in the 1940s, which involved him going to Black high schools and asking teachers about their top mathematics students. When colleagues at Morgan said the top students would only be frustrated by the lack of opportunity and employment they would eventually experience due to their being Black, Stephens replied, "Nonsense. If you're educated, you're educated and that's a good thing."

This one-person strategy, with the support of colleagues, evolved into a successful undergraduate mathematics program at a small Black college in rigidly segregated Baltimore, Maryland, in the 1950s and 1960s. This work encompassed not only ensuring that Stephens's students learned rigorous mathematics content that would prepare them for graduate study in mathematics—and Stephens was very clear about the end goal—but also provided a kind of orientation to how mathematics as a profession was done. As Earl Barnes related:

> When I got to Morgan, there was Dr. Stephens, who has won all kinds of awards for his teaching. He made a point of identifying those students who seemed interested in math and had a little mathematical ability and nurtured that ability. During my freshman year he selected a small group of us for special treatment. He put us in something like what we call today an REU [research experiences for undergraduates] program. Students in the program were paid a stipend to work on various projects in mathematics, and the projects were designed to teach us how to read mathematics with understanding. My assignment was to master the first three chapters of *Real Analysis* by McShane and Botts and to write a report summarizing what I had learned. That was a tough assignment for a first-year college student, and my report probably wasn't very good, but I learned a lot of mathematics trying to carry out that assignment. Reading that text forced me to seek out other books on real analysis. I didn't realize it at the time but Stephens was really preparing his REU students for graduate school. And on that count he was very successful. We all had fairly strong backgrounds in real analysis and abstract algebra by the time we reached graduate school. Incidentally, my stipend for that summer in 1965 was $600, quite a lot by today's standards.

These three mathematicians, two of whom returned to Morgan State University as faculty members after careers in other institutions, credit Dr. Stephens for their success:

> [At Morgan State, Stephens] made the effort to try and get as many talented people in math together [as he could]. We were in one of these calculus courses just for math majors. He came in the first day and he had a picture that he put on the bulletin board, a picture of a nice building. It was the Institute for Advanced Study at Princeton. That was the picture that he put up. And then he told us, "You are to aim for here; to get here at the Princeton Institute." You know, where Einstein was. He said, "Even if you miss, at a minimum, you

> will have a PhD. Because to get there, you would have to have a PhD, so if you aim here, you can miss." So in other words, establish that and set your goals very high in aiming that so that if you don't make it, you will be pretty well off. (A. Grainger, personal communication, 2008)

> Stephens's program was really, I think, rather innovative. First of all, we students worked with each other a lot. We even tutored students who weren't as good as [we were]. We worked with each other a lot, and then we found ourselves competing for things. But even from our sophomore year, he'd bring in a copy of the *American Mathematical Monthly. Mathematical Monthly*, which had two types of programs in it, advanced problems and elementary problems. Supposedly, elementary problems were solvable by senior math majors and early grad students. But the advanced problems were more for serious mathematicians, professional mathematicians. But he would bring these problems in and we would just work, trying to solve all of them. (S. Williams, personal communication, 2008)

Stephens's program, which he replicated in his work at SUNY Geneseo and Potsdam, has been chronicled in various articles and a book. Stephens himself has also left a record, in his own words, about the work at Morgan State and his teaching philosophy. In an interim report to the National Science Foundation in 1962, which funded the summer program that Williams, Grainger, and Barnes participated in, he wrote that "many prospective good mathematics majors become victims of the many social activities during the regular academic year. The summer program eliminated this temptation and provided a better atmosphere for concentrated study." The care with which Stephens and his colleagues selected students for participation, developed the program activities (which included seminars in topology and algebra, reading and analysis of research papers and texts, and solving problems from the *American Mathematical Monthly*), assigned students tasks and leadership responsibilities, and evaluated their progress is evident. In addition, Stephens hired recent Morgan State graduates, who were themselves pursuing graduate studies in mathematics, as paid, "regular" supervisors of the program. Assistant supervisors, who were mathematics faculty at other institutions, were unpaid. Stephens, the director of the project, included himself in this category.

Although Stephens left Morgan in 1962 before their graduation, Barnes, Grainger, and Williams went on to graduate school and professional careers as mathematicians. They credit Stephens's rigorous program and mentoring for their success—as Art Grainger notes:

> [Doing well in mathematics is partly due to] your individual ability, but it's also a combination of factors. If you can get in an environment which we were in, just that interaction even magnifies [your ability] more. . . . The idea is that [we succeeded] because of the program and having that environment of mentoring and because of Dr. Stephens, who realized that and worked hard to get that particular environment.

Stephens's former students and their colleagues at Morgan State have developed a "Stephens Scholars" program for high school students to attract them to mathematics. What Stephens's program showed—and what he later replicated at predominantly White institutions—was that students could learn mathematics at a very high level. He exposed students to graduate-level content during their undergraduate years and provided a space in which they could learn with and from each other. One respondent noted that Stephens's extensive preparation meant that the "first year of graduate school was a breeze." Two of those students, with GRE scores above the 90th percentile, were among the first to desegregate the graduate program in mathematics at the University of Maryland.

Like the Morgan State alumni in the previous section, other Black mathematicians who attended Black colleges in the 1960s describe faculty in the sciences at these institutions as integral to their success. But younger mathematicians also describe similar relationships with faculty and peers. I describe similar practices in a more contemporary context by discussing Spelman College, which has been recognized as having an effective undergraduate mathematics program (Tucker, 1996).

As a physical space, the Albro-Falconer-Manley Building, also known as the "Science Building," on Spelman's campus has immediate visual impact. The Science Building is named for three women in the sciences (two of whom are Black Americans) with influential careers at Spelman: Helen Albro, Etta Falconer, and Audrey Forbes Manley. When one enters the building, the first thing one sees is an atrium with museum display cases, photographs, and other artifacts documenting the impressive history of women in the sciences at Spelman.

For Black women mathematicians in particular, including five Spelman alumni and professors who were interviewed for this study, Etta Falconer

(1933–2002), who earned her PhD in mathematics in 1969 from Emory University, looms large over the science and mathematics programs at Spelman. Her influence is not limited to Spelman students—a male mathematician who attended Morehouse College also describes her impact and credits conversations with Dr. Falconer with his choice of doctoral program:

> Oh, she was so wonderful, I would just go stop by, you know. I used to like to go to Spelman and just visit. I would periodically just walk in, just pop in her office and whenever she could I'd take a seat and she'd ask me how things are going, you know, we'd just talk a little. And I didn't understand how busy professors are, I just didn't. I had no idea so anytime someone comes knocking on my door I try to make a little time for them. If not on the spot, I'll tell them, you know, "Come back at 4:30" or whatever. But I always try because I think about how she was generous with her time and her wisdom.

Several mathematicians who are now college faculty speak of Dr. Falconer's approach to ensuring that students and alumni of Spelman were successful in their graduate science programs. One, Sylvia Bozeman, now a long-time faculty member and administrator at Spelman, began her career as a mathematics instructor there and describes Dr. Falconer's support for her professional career and for the completion of her PhD:

> I had the best mentor for 30 years here at Spelman, Dr. Etta Falconer, and she was just incomparable as a mentor. I came to Spelman really inexperienced, having recently completed a master's degree. I learned so much from her.
>
> She had just the most wonderful devoted following among all of us here in the sciences, because she didn't just mentor me, she mentored lots of faculty, staff, and many, many students. There were lines of students outside her door all the time. Even her administrative assistants and secretaries were influenced by her and encouraged to return to school and obtain higher degrees—you know, they all went to school and left her. They all moved up. She mentored everybody.

A Spelman graduate who herself earned her PhD after 2003 was mentored both by Falconer and Bozeman and, like others, attributes Spelman's success in the sciences to Etta Falconer and her colleagues' lasting influence:

Etta Falconer mentored me. I mean, she was really responsible for this sort of renaissance of science at Spelman.

And she was old school, she just kind of told you what to do, like, "You're going to present here," and, "You're going to do this," and, "You're wrong, you're right." So it wasn't a nurturing kind of mentoring, she just kind of told it.

These relationships continued long after students left Spelman:

Well, Etta Falconer, I appreciated the fact that she gave me chances when I was a graduate student to come back and talk to the undergraduate majors about what I was doing. And on several occasions I've been able to talk to the majors at Spelman. As I would run across her at professional meetings, she would always have good advice about things in my career. It was always good to talk to her, to hear her wisdom and just to listen to her talk about any particular subject was always special to me.

The mathematics program at Spelman is rigorous and effective, thanks to its tremendously committed faculty, but its success is at least partially due to the extensive mathematics socialization that its students experience at a very high level.

Spelman's pre-college program in math and science was really important to me. We were able to sort of get a leg up coming in. But the biggest thing was meeting these other scientists, other Black women interested in math and science and just studying, that social dynamic was just huge, huge, huge. That was great.

Another Spelman graduate who is now a mathematician noted:

Spelman was the first time I saw Black women with PhDs in math, which I think has a very interesting impact. I think I only recognized it in hindsight because when I was there I sort of took it for granted. We had about five or six Black women on faculty that had PhDs, and I just thought, "Oh, they're a dime a dozen."

And it wasn't until my junior year that I realized there were fewer than 100 Black women who had ever gotten PhDs in math. Not only that, to know that if I got one in five years I would still be in the top 100. It was just blasphemy.

Spelman alumni and faculty interviewed for this study speak a great deal about the mathematics community at Spelman—how students work together and collaborate, and how the faculty structure mathematics opportunities that are accessible to students. As one participant noted:

> It was easy to major in math at Spelman because they're supportive. You know, we did our homework together. Nobody was like, you know, "I want to be the top student, and I won't help you or share it with you."

The preceding discussions reveal that Spelman and Morgan State (and several other Black colleges and universities described by mathematicians in this study) have operated as sites that contribute to Blacks' mathematical knowledge and mathematics identity development. In addition, these institutions have also contributed substantially to their graduates' induction into the mathematics profession. But it should be acknowledged that Black colleges and universities have made contributions beyond the development of strong mathematics majors. Their boundaries are permeable and expansive: their practices, disseminated through faculty and alumni, continue to influence graduates and the development of networks that facilitate development of mathematical talent, not just for Black Americans. For example, Sylvia Bozeman and her colleague Rhonda Hughes of Bryn Mawr designed and developed EDGE (Enhancing Diversity in Graduate Education) to attract and retain talented women in graduate study in the mathematics sciences (Bozeman & Hughes, 2004). This program has inspired several versions of similar programs, some targeting diverse students in general, regardless of gender.

Although it is not the case that Black colleges alone can produce Black mathematicians—far from it—it is noteworthy that the most successful undergraduate and graduate programs owe more than a small debt to the structures and practices of Black colleges' mathematics programs (Cooper, 2000, 2004). In many cases this is accidental, but in some others, the programs are explicitly built on effective practices of Black colleges. Stephens's Potsdam and Geneseo success after he left Morgan State is but one example. And a number of Black mathematicians—regardless of whether or not they attended an HBCU—have become administrators of mathematics departments at predominantly White institutions that successfully produce Black mathematicians. For example, Raymond Johnson, the mathematician who was one of the first Black graduate students to desegregate Rice University when it was still chartered as "an institution for the White Citizens of

Texas," helped to facilitate the University of Maryland's success in attracting and retaining Black students in graduate-level mathematics when he was its mathematics department's chair. (It made the national news when three Black women graduated with PhDs in mathematics from the University of Maryland in the same year). Johnson has described his work at the University of Maryland as facilitating professional networks for Black students and ensuring that they collaborated with each other as well as joining networks of other students. Although Johnson is modest about his efforts, Cooper (2000) states that many of the Black graduate students at Maryland "cite[d] Johnson as a reason they elected to come to Maryland for the PhD, having met him in recruitment settings like Undergraduate MATHFest, an annual conference sponsored by the National Association of Mathematicians designed to inform Black mathematics majors about . . . graduate and career opportunities in mathematics" (p.182).

Conclusions: Mathematical Spaces and the Importance of Mathematical Meanings and Histories

Black colleges as mathematical spaces have important social and cultural meanings in the lives of Black mathematicians. They serve in many ways as a rebuke to historic notions of White intellectual superiority, especially in the South, and create opportunities for Blacks to demonstrate mathematics excellence. Mathematics opportunities within Black colleges are largely unconstrained by coded discussions of intelligence and merit that may arise in predominantly White settings in which Blacks are a distinct minority. Indeed, a recent PhD graduate noted that at a Black college, "no one is surprised that you do well in math." Even fictional Black colleges can have a profound impact, as a young Black mathematician attests:

> So, you know, we had grown up with *A Different World*. I was in love with Dwayne Wayne. And I thought I was so cool. Like, "Ooh. It's cool that he majored in math." I had this glorified image of going to an HBCU and majoring in math. I thought I was cool like Dwayne Wayne. Part of me was just like, "I guess that will be okay," because I saw it. And I was just like, "Okay. Well, I saw it on TV so it's not crazy."

Further, Black colleges' physical and visual attributes serve important purposes. The artifacts displayed at some Black colleges, at Spelman, for example, underscore that Blacks belong in mathematics and science fields. In a world where books about mathematicians can include none or very few Black American mathematicians, these kinds of visual artifacts and shared histories are critical (see Kenschaft, 2005, as an exception). Virtual spaces, such as Dr. Scott Williams's (a Morgan State College alumnus) website *Mathematicians of the African Diaspora* (http://www.math.buffalo.edu/mad/index.html), are also contributing to a more widely known existence of Black excellence in mathematics.

Further, the networks and communities Black colleges directly and indirectly influence—Black mathematicians' organizations and successful mathematics graduate programs for Black students within predominantly White institutions—adopt these meanings. But what contributes most greatly to the successful work of Black colleges, Black mathematicians' organizations, and certain graduate programs as incubators for Black mathematical talent is the presence of key individuals—Black and non-Black—who are committed to developing mathematical talent and excellence. Much of the research describing factors that contribute to the high achievement of underserved students in mathematics, in particular, points to the importance of relationships that have personal dimensions and relate to the content (Berry, 2008). It is certainly true that, undoubtedly, mathematicians belonging to other demographic groups may experience similar mentoring and support, but the narratives of Black mathematicians suggest that high expectations for their success and access to mentoring and networking are things they feel they cannot take for granted. Further, because most Black mathematicians earn their graduate degrees and have postgraduate experiences in institutions and settings in which they are the sole Black mathematician, this context permeates their professional experiences. And as the narratives of mathematicians reveal, these racially coded experiences are not limited to older mathematicians. Thus Black colleges, as well as Black mathematicians' organizations, serve as spaces that continue to provide important support for talent development. They continue to serve as models and inspiration for younger generations of mathematicians and for initiatives and institutions that are committed to developing mathematicians and scientists.

SIX

CONCLUSIONS

The mathematical life histories of Black mathematicians reveal much about how they come to know and do mathematics in multiple contexts—home, school, community, college, university, and the profession. Although these mathematicians have varied paths to the profession, they share a common social and cultural bond—they are African Americans in the United States.

This is not to suggest that all African Americans, or all African American mathematicians, have exactly the same experiences; however, there are significant commonalities within and across generations of mathematicians. For example, older mathematicians who entered high school, college, and graduate school before the zenith of the Civil Rights Movement have varied experiences in those institutions: Evelyn Boyd Granville grew up in segregated Washington, DC, and attended Smith and Yale, two northern, predominantly White and elite institutions, but she states that she "really had no racial problems there." David Blackwell attended predominantly White institutions for high school and college and university in Illinois, and he describes matter-of-factly the segregation that was imposed on Black students there at that time. Clarence Stephens attended rural segregated schools in the South, including a Black boarding school, because there were no public secondary schools for Blacks at that time in that state; a historically Black college; and the University of Michigan for graduate school. What all these mathematicians have in common is the knowledge that their Blackness precluded them from being able to demonstrate their excellence as faculty members at White postsecondary institutions. Eventually they all worked at predominantly White institutions, but this occurred some time after they received their doctorates.

Younger Black mathematicians find themselves, still, at a curious intersection of race and opportunity. Although opportunities are much more widely available for them than for those who came before them, they are still subject to sometimes hidden and sometimes stunningly explicit discourses about race and merit. Throughout their mathematical journeys, they may find themselves isolated in high school and college mathematics classrooms, if they are admitted to them at all, and they may have to respond to queries about their mathematical preparedness and expertise in graduate school classrooms. They may find themselves members of a very small group of Black students in classrooms, high schools, and colleges. They ponder what happens to other students, who may be equally talented but whose parents may lack the advocacy skills needed to ensure placement in advanced mathematics classes or who accept the opinion of professors that perhaps the mathematics major is not for them.

The mathematics trajectories of these mathematicians, for the most part, follow a traditional path of demonstrating excellence in high school, becoming a math major in college, and pursuing and succeeding in graduate studies. However, there are enough exceptions to this path to make us consider (and recognize) that it is quite possible that we could have more mathematical scientists—of any racial background, but certainly of African American heritage—than we currently do. Within the group in this book, there are mathematicians who did well in school mathematics but were disinterested in pursuing it until they had an experience with a charismatic and/or knowledgeable, excellent teacher; there are mathematicians who struggled in other high school courses but excelled in mathematics, yet were in danger of being overlooked due to their grades in those other courses; there are mathematicians who struggled with mathematics in college but still exhibited a keen interest in the subject; there are mathematicians who majored in other subjects and discovered, or rediscovered, their love for mathematics much later than the traditional college student; there are mathematicians who pursued other careers altogether (in the military, as activists, as high school teachers) before deciding to pursue graduate study in mathematics; there are mathematicians whose graduate school experiences were less than optimal but who still went on to succeed at subsequent institutions. At each of these critical transition points, there were gateways back to mathematics that were provided by parents, teachers, or college and university faculty. Some of these moments could only be called serendipitous—in that there was great potential for these mathematicians to be overlooked as strong, or potentially strong, mathematics students. These doors could easily have remained closed. That a number of mathematicians share stories about parents having to advocate strongly for their placement in high school advanced mathematics courses and stories about being discouraged from pursuing mathematics as a major or as a discipline for graduate study should give us great pause.

Aside from the potential pitfalls of these numerous critical transition points, there are some obvious examples of practices and policies that promote Black mathematicians' success. A common refrain in these narratives is that of the importance of Sputnik, not so much its launch but rather the ramifications of the Russians being perceived as winning the space race. The desire to trump the Russians led to an unprecedented U.S. investment in science, technology, engineering, and mathematics education at all levels. Increased funding for multiple types of programs for science, technology, engineering, and mathematics (STEM) benefited many second-generation mathematicians in this study. Mathematicians who were excellent high school students in math attended summer and Saturday STEM programs at universities and colleges; mathematics majors in college participated in summer research experiences funded by the National Science Foundation and were hired by faculty as research and teaching assistants under the auspices of professional development programs for minority mathematicians. Without this funding stream, it is possible that this set of mathematicians might not have had this exposure during that time period, when segregation was still prevalent. Third-generation mathematicians also benefited from significant funding from NSF, attending and participating in programs such as those described above, as well as initiatives designed to address the issues facing diverse students in graduate mathematics programs, including programs cofounded and designed by Black mathematicians. In addition, Black mathematicians seek and secure funding for programs targeting not only college students but also high school students, to attract them to mathematics. These varied and critical experiences speak to the value of continued and increased funding for such programs.

The necessity of these experiences becomes increasingly clear when one considers the fact that many mathematicians in high school and college were largely unaware of the opportunities that their mathematics talent could afford them. Some were largely unaware that they could go to graduate school for mathematics until quite late in their undergraduate college years. Quite a few "backed into" mathematics careers, with many assuming that they would be high school teachers, or else, advised by well-meaning teachers or adults, they were steered into engineering. There is nothing wrong with these careers, but for some mathematicians it was somewhat by chance that they happened to discover opportunities for graduate mathematics programs and that, in reality, they preferred the study of mathematics to engineering. For some, there were important avenues of exposure in secondary school, through science and math clubs and through enrichment activities to which they were directed by teachers and administrators. A few others had parents with advanced degrees in the sciences, who made them aware of opportunities in mathematics. But for most, lack of exposure to what mathematics careers might be all about was a common refrain.

Somewhat surprising in this study is that in reality, generational differences in terms of mathematics experiences were limited to the state-sanctioned discrimination that was in evidence during mathematicians' secondary and college schooling years. Black mathematicians of all three generations attended predominantly Black and predominantly White secondary schools, as well as historically Black and predominantly White institutions for college. Although a few Black mathematicians of the second generation reported being actively involved in civil rights and war protests in the 1960s and 1970s, this was not the norm for mathematicians of their generation or, for that matter, for those of the first or third generations. Black women mathematicians, unlike their White female counterparts, for the most part report little challenge to their mathematics pursuit until they arrived in graduate school mathematics programs.

The narratives of mathematicians have critical implications for mathematics teaching and learning in schools and communities. Indeed, early in their mathematical life spans, it is clear that much mathematics learning and socialization occurs outside of formal school settings. From their earliest mathematical memories, when they were engaging in family traditions and exposed to mathematics concepts by family members and other adults, to teachers going above and beyond the formal classroom parameters to introduce mathematicians to advanced material, to their participation in summer, after-school, or Saturday mathematics enrichment programs, Black mathematicians in this study reported significant exposure to mathematics. Noteworthy is that the family members (immediate and extended) of Black mathematicians who exposed them to these mathematics concepts range from those who never completed high school to PhD holders themselves. This was true across generations.

When Black mathematicians described their formal high school mathematics experiences, most of these stories sounded very similar to what contemporary students report. In short, mathematics class is often a series of mind-numbingly boring exercises to get through. Although many reported that they were always good in math and could complete these exercises easily, some mathematicians reported vivid and fond memories of mathematics class experiences that did not fit this mold. When they encountered teachers who showed that mathematics was engaging and interesting in and of itself as well as useful, and that mathematics was not always about simply generating one right answer, this was often a watershed moment in their mathematics development. Several described acquiring new "habits of mind" regarding mathematics on the basis of these experiences: they realized that one could prove things using mathematics, that mathematics was something to be thought about and appreciated, that mathematics was not just about "plugging and chugging."

Many Black mathematicians who attended historically Black colleges and universities found communities of students and faculty who were invested in their success and had high expectations for their continued study of mathematics. The most successful Black colleges in which departments of mathematics can be characterized as mathematical spaces are *intentional*, and as such, are spaces that are structured in deliberate ways to foster rigorous mathematical knowledge development, socialization and identity building, and induction into a community of mathematics doers (in this case, the mathematics profession). Of particular interest is that most Black colleges in and of themselves have little formal responsibility for graduate training—their graduates, for the most part, earn their PhDs from predominantly White institutions. But Black colleges take the long view: they develop networks that persist long after students graduate, and the strong relationships between undergraduate faculty and alumni continue throughout their professional careers (Cooper, 2004). Black mathematicians' organizations in some ways replicate the engaged and affirming environment of Black colleges for graduate students as well as for junior and senior professionals.

Thus, for Black mathematicians who feel a strong mission to develop other Black mathematicians, their own formative and educational experiences play a significant role in how they do this work. Further, for those who attended historically Black colleges and universities—and for those who participate actively in Black mathematicians' organizations—the traditions of those entities facilitate their effectiveness at enacting practices that support the development of Black mathematical talent. They view their professional identities as encompassing not just their research and teaching but also a "mission" to attract others to the field and the desire to "pass things on." They draw on their experiences in schools and communities to formally or informally craft intentional spaces, rather than allowing for (or hoping for the possibility of) inadvertent spaces, in which young people learn and practice mathematics, develop strong mathematics identities themselves, and are inducted into a community of mathematics doers.

The mathematical spaces crafted and inhabited by Black mathematicians have important social and cultural meanings in their lives—they are not culture free or color-blind. They exist as sites of resistance, particularly in the case of historically Black colleges and universities and Black mathematicians' organizations, as a rebuke to historic notions of White intellectual superiority in mathematics. They also exist, particularly within school and university classroom contexts, as sites of isolation and belongingness, as well as sites that present opportunities for and obstacles to mathematics success. As Black mathematicians describe them, these dualities cannot be neatly partitioned into predominantly Black versus predominantly White settings.

The practices within affirmative mathematical spaces for Black students are informed by historical memory, the social and cultural experiences of Black mathematicians, and by the desire of many Black mathematicians to challenge the portrayal of mathematics as solely a White endeavor and the myth of Black inferiority in mathematics. In crafting mathematical spaces that support Black students'—as well as other underrepresented students'—mathematics learning, socialization, and performance, Black mathematicians draw on a reservoir of cultural resources that have largely gone unexamined in much of the literature concerning Blacks and mathematics in postsecondary settings.

These findings have implications for schools that serve African American students and others underrepresented in the sciences. School practitioners need to be aware of the fact that even when Black students exhibit high achievement in mathematics, they may continue to be positioned as low-achieving despite their excellence, and this has implications for how they are granted access, or deemed worthy of access to, mathematics resources. This is not a new story—as Wiggan (2007) states, when a researcher conducted a study in the late 19th century and Black students in Washington, DC, outperformed their White counterparts on an examination, the researcher "asserted that Black students were intellectually deficient" and critics of the test "called for a revision . . . because the outcome did not support the prevailing belief in White superiority" (p. 312). For too many of the mathematicians in this study, there are some significant threats to opportunity. Sometimes the identity that had been ascribed to a participant by others could have interfered, or did interfere, with his or her opportunities to do mathematics and join the pipeline of courses and experiences needed in order to become a mathematician.

But, importantly, these findings have implications for how we think about out-of-school experiences that cultivate mathematics identity and potential. Mathematicians' narratives echo previous findings in a study with high-achieving high school students (Walker, 2006) that show that persons and relationships from multiple worlds formed academic communities (comprising family members, peers, teachers, and others) that had an important impact on mathematics learning and socialization. In addition, as I discovered in the high-achievers study, the persons providing support or socialization opportunities for the mathematicians are not necessarily themselves mathematics teachers or mathematicians. In fact, some of these persons are those who would be considered "uneducated" in the formal sense by many.

These narratives suggest that we could be much more successful in improving mathematics outcomes and fostering interest in mathematics. We have to rethink how and where mathematics learning and practice occur and where one's mathematics identity is developed. To do this, we should build on out-of-school spaces that support mathematics socialization and also reimagine the mathematics classroom as a space that supports mathematics identity development and positive socialization experiences as well as one that provides opportunities to learn meaningful mathematics (Walker, 2012b).

But we also have to think about how we ensure that meaningful mathematics occurs beyond fleeting conversations, students' individual experiences and identities, and the spaces in which students happen to find themselves. Our expectations of students' abilities are key—if we think students have potential and if they are worthy of our attention in nontraditional spaces that support mathematics learning, we are likely to become much more intentional and purposeful about creating these spaces. Thus, opportunities to engage in meaningful mathematics have to have intentionality and purpose and should not solely be haphazard or happenstance. For too many of our students, particularly our underserved Black students, these opportunities are limited.

Black mathematicians across generations have issued a call to action. And although they are convinced that mathematics as a discipline is a worthwhile endeavor, they acknowledge that practices and structures within the broader mathematics community do not necessarily support or invite Blacks into the field. However, they craft environments that facilitate the development of mathematics excellence for those who find existing structures unwelcome.

The spaces that they have experienced during their mathematical lives have encompassed both informal and formal learning opportunities; occupied a plethora of physical locations, including neighborhoods, schools, classrooms, and homes; and ranged from official, institutionally sanctioned activities led by trained professionals to fleeting and happenstance meetings that spurred reflection and motivated mathematical interest and pursuit. The spaces they craft are generative, rooted in history and memory, and comprise practices that reflect their own experiences and are deliberately fashioned to facilitate their own success and that of others.

Their recollections should underscore that the pipeline to mathematics does not begin in college or graduate school, and although much of the literature regarding student achievement reifies limiting and narrowing discourse about Black students, there is substantial evidence that many Black high school students find the mathematics classroom an unwelcoming and uninspiring place. Wade Ellis, Jr.'s words in 1980 still, sadly, hold true today:

> Nowadays our promising youth are even more menacingly threatened by exposure to teachers . . . who have been convinced to their very viscera that Blacks . . . are abysmally and irrevocably hopeless as far as mathematics is concerned. (Newell et al., 1980, p. ix)

It is critical that we continue to conduct work that seeks to challenge dominant discourses that contribute to notions of inferiority in mathematics for Black children. The narratives of these Black mathematicians, in their complexity, suggest not a one-size-fits-all paradigm, a single strategy to foster excellence, or even a single location for mathematics learning; they suggest that we draw upon young people's extensive social and cultural resources to better inform both in and out of school practice and ameliorate inequities and disparities in opportunities and resources that inhibit excellence.

Much more work exploring these issues needs to be conducted. Although this book paints a unique portrait of Black mathematicians and their formative, educational, and professional experiences, there are many more stories to be told and histories to be documented. Originally, this study was to be a relatively straightforward endeavor documenting Black mathematicians' first mathematical memories and school experiences. I was particularly interested in documenting the school experiences of those mathematicians who attended segregated schools in the South and were among the first to desegregate their high schools, colleges, and graduate schools. It soon became clear that there were multiple interlocking stories to be told, some of which I have touched on in this book and others that remain, sadly, untold. There are stories that emerge from these narratives that might never have been known, and these stories, as tantalizing ephemera, reveal the importance of documenting the lives and experiences of our elders. Just some of the stories begging to be explored much more deeply are those of segregated boarding schools in the South, such as the Harbison Institute that Clarence Stephens attended, which operated within a very particular social, cultural, and historical context, in a time when Blacks found it nearly impossible to get state or local legislatures to fund public Black high schools in the segregated South. Documenting the curricula and mathematics practices of these schools—including the Dunbar School in Washington, DC—and many others is well worth doing, as is conducting histories of key mathematics departments in historically Black colleges and universities (Howard, Fisk, Morehouse, Morgan State, and Spelman, for example). In addition, exploring the pipelines between historically Black colleges and universities and White graduate programs—notably, the Howard/University of Pennsylvania connection comes to mind, as well as the programs that were among the first to admit Black students to their graduate programs—Iowa State, Michigan, and others. A much more in-depth exploration of Black professional organizations for the sciences, beginning with the undersung Beta Kappa Chi, is also critical to undertake. Archives and records in historically Black colleges and universities and predominantly White institutions alike could contain critical information about all of these matters. It would be shameful if these histories just lived in people's memories and not the printed word. Any and all of these potential avenues of exploration are undoubtedly book-length projects, but a start would be to engage education researchers, historians of science, social scientists, and graduate students considering dissertation topics to take on these projects.

These hidden stories, and those of mathematicians such as Benjamin Banneker, Thomas Fuller, Elbert Cox, Euphemia Lofton Haynes, and others, should make us wonder what we have missed learning about how people come to know, love, and do mathematics and, further, what we can learn from the effective and essential practices of institutions that supported their development. There are some who argue, strongly, that young people who exhibit keen mathematics talent will always be discovered and supported. The sheer number of potential missed opportunities to pursue and persist in mathematics for many of the mathematicians in this study should counter that argument. Although it is undoubtedly true that neither Benjamin Banneker nor Thomas Fuller could have dreamed of a world in which there is a significant number of Black mathematicians, it is up to us to ensure that we continue to identify and develop mathematical talent, wherever we find it, and that these stories are shared, preserved, and not forgotten.

APPENDIX A

METHODOLOGICAL NOTE

Qualitative research methods were used to carry out this study of Black mathematicians' formative, educational, and professional experiences; in particular, narrative research methods were used. Narrative research seeks to allow participants to author their own stories about past and present experiences, and the use of story as a frame to reflect upon, understand, and share these experiences, particularly in the sciences and education, has been used by many researchers (e.g., Burton, 1999; Moore Mensah, 2009; Stage & Maple, 1996). In particular, the use of narratives to understand and examine the spaces that Black mathematicians have inhabited and continue to inhabit, given their social, cultural, and political realities, allows the researcher and readers to gain a complex, nuanced understanding of how Black mathematicians have operated and continue to operate in multiple contexts. A qualitative inquiry such as narrative research allows for a deeper investigation of the factors that facilitate and impede Black progress in mathematics and contributes to the building of theory about mathematics learning, socialization, and performance.

Participants

Thirty-five African American mathematicians were interviewed. All were born in the United States, and their PhDs, in mathematics or a mathematical science, were granted in three key time periods: between 1940 and 1965, between 1965 and 1985, and from 1985 through 2010. I have chosen to purposefully select individuals in this way, as did Pearson (1988), because of the contextual history and realities of Black education in the United States, which change over time (O'Connor, Lewis, & Mueller, 2007). Mathematicians who earned their PhDs before 1965, whom I refer to as "first-generation" mathematicians, faced overt discrimination and racism that was enshrined in the laws of the United States—there were both *de jure* and *de facto* restrictions to their securing an education and a career. Mathematicians who earned their PhDs after 1965 but before 1985—"second generation"—benefitted from the dismantling of legal obstacles to civil rights and education rights but experienced state-sanctioned segregation for a period when they were attending elementary and/or secondary school, as well as in daily life. Mathematicians who have earned their PhDs after 1985 form a younger cohort, the "third generation," which has, in theory, not experienced legal segregation. In addition to these contextual factors, this partitioning of time allows the examination of narratives across roughly equally 20-year increments.

Of the 35 mathematicians interviewed for this study, 17 attended predominantly White institutions for their undergraduate education and 18 attended historically Black colleges or universities. All except two attended predominantly White graduate schools for their PhDs. The majority are now employed by colleges and universities as faculty members and administrators; several are retired from university careers. Six have worked for private industry, research organizations, or the federal government.

Data Collection and Analysis

Interviews were conducted using a semistructured, open-ended interview protocol developed by the author (Walker, 2009) and lasted between 45 minutes and 3 hours. Most interviews lasted at least an hour. Interview questions for the most part were arranged chronologically, to allow the participants to reflect on their mathematics experiences during early childhood, adolescence, college, graduate school, and their careers. The questions were designed to be sufficiently open ended to allow stories to emerge, and follow-up questions were determined according to the direction of the interview. The first question, which asked them to share their earliest mathematical memories, allowed mathematicians to define for themselves when they first became aware of mathematics and to describe their positive (or negative) response to that experience. Subsequently, in response to questions, mathematicians shared, for example, stories about doing mathematics at home or in the neighborhood with family members and neighbors; about significant classroom experiences and out of classroom experiences with youth and adults that supported their mathematics learning; about their experiences at mathematical conferences and in mathematics organizations; and about their professional experiences in university, government, and industry careers. Mathematicians' descriptions of institutions and organizations were cross-checked, where possible, with other mathematicians' interviews and recollections, documents about those institutions and organizations, and follow-up conversations with informed and expert colleagues associated with institutions and organizations. Interviews were transcribed, read several times, and coded by the author and a research assistant initially for major themes.

Since 2008, I have been a participant-observer at several mathematics conferences and special meetings involving Black mathematicians, including CAARMS (2008), NAM (2009), and EDGE (2009), as well as at conferences and meetings focusing on diversity in mathematics hosted by various institutions and organizations, including the National Security Agency (NSA) and the Mathematical Sciences Research Institute (MSRI). I took extensive field notes at each meeting or conference, focusing on discourse about educational and professional experiences as well as on mentoring activities and discussions that occurred. At some of these conferences, I was an invited speaker. In addition, I have visited several historically Black colleges and universities, observing mathematics classes and meetings, conducting interviews with faculty, and giving talks.

Although information about formative experiences and historical aspects of Black mathematicians' development is somewhat limited, I have analyzed historical documents relating to Black mathematicians. These include but are not limited to archival records; oral histories; biographies; and books and newspaper, magazine, and journal articles relating to Black mathematicians (e.g., Agwu, Smith, & Barry, 2003; Bedini, 1999; Cooper, 2004; Datta, 1993; Dean, 1997, Kenschaft, 2005; Newell et al., 1980).

The interview transcripts were used to develop common narratives about Black mathematicians' formative, educational, and professional experiences. A research assistant and I examined the interview transcripts for narratives that spoke to experiences that related to mathematics teaching and learning (within and outside of schools), the development of mathematical identity, and mathematical spaces. We identified major themes relating to these ideas that incorporated racial, social, historical, cultural, political, and physical aspects. For example, building on Martin's definitions (2000, 2006) of identity, these major themes included dispositions and beliefs about mathematics, mathematics activities and practices, and descriptions of self in relation to mathematics (and whether those were authored by oneself and/or others).

APPENDIX B

INTERVIEW PROTOCOL

Part 1.

Please describe your early childhood experiences with mathematics. How did you become interested in mathematics?

Were any family members mathematically inclined? Please explain.

Please tell me about an experience from your [elementary/secondary] school days that was important to your mathematical thinking.

Did your teachers in elementary/secondary school support your mathematics learning? If so, how?

Did you talk with your friends about mathematics? Did you participate in math clubs, etc. in elementary or secondary school? What kinds of mathematics experiences did you have with friends in and out of school?

Part 2.

For your undergraduate education, did you attend an historically Black college or university (HBCU) or a predominantly White institution (PWI)? Please describe your mathematics class experiences at your undergraduate institution.

How did you decide to major in mathematics in college? What was the response of your family/friends/teachers to that decision?

What kinds of out-of-classroom activities did you participate in [in] college that supported your mathematics learning?

Did you talk with your college friends about mathematics? What kinds of mathematics experiences did you have with friends in and out of school?

Part 3.

How did you decide to continue your education to get the PhD in mathematics? What factors contributed to that decision?

Who were your mentors? How did they mentor you?

Have you sponsored any PhD students yourself? How is your relationship with them similar/different to the relationship you had to your PhD sponsor? Why is it similar/different?

What would you describe as your priorities as a mathematician? (probe: research/teaching/publishing/mentoring/)

What kinds of mathematical experiences do you engage in with family members [nieces/nephews/children/grandchildren]? How are these similar or different from experiences you had as a child?

Are any of your family members . . .

- participating in mathematics activities in secondary school?
- majoring in mathematics in college?
- planning to pursue the PhD?

NOTES

Preface

1. From the definitive biography of Benjamin Banneker by Silvio Bedini, initially published in 1972 and reissued in 1999.
2. Newell et al., 1980.

Chapter 1. Introduction

1. See Appendix A for a discussion of the methodological approaches used in this study.
2. All quotations in this book, except as otherwise stated, are from interviews conducted by the author between 2008 and 2010 at the interviewees' homes or offices, during conferences or meetings, or over the telephone, as described in the appendix. With the exception of quotations from the Vanguard and selected quotations from Sylvia Bozeman, Robert Bozeman, Raymond Johnson, Earl Barnes, Arthur Grainger, and Scott Williams, pseudonyms are used for the interviewees.
3. Browne finished the requirements for the doctorate in 1949, but it was not awarded until 1950.
4. Graduates of Dunbar (or M Street School, as it was first known) include luminaries such as Carter G. Woodson, Charles Drew, and Mary Church Terrell.
5. Dr. Lofton Haynes's personal and professional papers are archived and held by Catholic University in Washington, DC.

6. Adelaide Cromwell is professor emerita of sociology at Boston University and has written a family history, *Unveiled Voices, Unvarnished Memories: The Cromwell Family in Slavery and Segregation, 1692–1972*, published by the University of Missouri Press in 2006.
7. Recently, Morgan State University mathematics professors have been awarded funding from the National Science Foundation (NSF) and National Security Agency (NSA) to develop initiatives for Baltimore Public Schools students and attract them to mathematics. Students participating in the program will be known as "Stephens Fellows" in honor of Clarence F. Stephens.

Chapter 2. Kinship and Communities

1. Etta Zuber Falconer (1933–2002) was a long-time professor at Spelman College, where her mother had attended. She earned her PhD in mathematics in 1969 from Emory University. Dolan Falconer, her son, received bachelor's and master's degrees in nuclear engineering from Georgia Tech University and is the CEO of ScanTech Holdings.
2. Given the increasing rates of resegregation of public schools in the United States, with Black students largely attending predominantly Black schools, the implications of this finding are sobering.

Chapter 4. "Representing the Race"

1. Raymond Johnson has spent much of his career at the University of Maryland, including a period as its Mathematics Department chairperson, and is currently at Rice as the W. L. Moody Visiting Professor.
2. William Claytor was the third African American in the United States to earn his PhD in mathematics, from the University of Pennsylvania.

Chapter 5. Flying Home

1. The other two HBCUs with doctoral programs in mathematics are Delaware State, which offers a PhD in applied mathematics and mathematical physics, and Clark Atlanta University, which offers a PhD in industrial and computational mathematics.

2. Many of the first Blacks to earn their PhDs in mathematics earned their undergraduate degrees from historically Black institutions—Dudley Woodard (the second) earned his bachelor's degree from Wilberforce, and Marjorie Lee Browne (who was one of the first three women to earn her PhD in mathematics) earned hers from Howard University.
3. One of Dr. Granville's faculty colleagues was Dr. Lee Lorch.
4. The two students she taught who eventually earned their PhDs in mathematics were Vivienne Malone Mayes (1932–1995) and Etta Zuber Falconer (1933–2002).

REFERENCES

Agwu, N., Smith, L., & Barry, A. (2003). Dr. David Harold Blackwell: African American pioneer. *Mathematics Magazine, 76*(1), 3–14.

Albers, D. J., & Alexanderson, G. L. (1985). *Mathematical people: Profiles and interviews.* Boston: Birkhauser.

Albers, D. J., Alexanderson, G. L., & Reid, C. (1990). *More mathematical people: Contemporary conversations.* Boston: Harcourt Brace Jovanovich.

Anderson, J. (1988). *The education of Blacks in the South, 1860–1935.* Chapel Hill: University of North Carolina Press.

Bedini, S. (1999). *The life of Benjamin Banneker* (2nd ed.). New York: Scribner.

Berry, R. Q. (2008). Access to upper-level mathematics: The stories of successful African American middle school boys. *Journal for Research in Mathematics Education, 39*(5), 464–488.

Black colleges and universities are graduating an increasing share of African Americans who earn Ph.D.s in mathematics and science. (2008). *Journal of Blacks in Higher Education, 61*, 35.

Boaler, J., & Greeno, J. (2000). Identity, agency, and knowing in mathematical worlds. In J. Boaler (Ed.), *Multiple perspectives on mathematics teaching and learning* (pp. 171–200). Stamford, CT: Ablex.

Bozeman, S. T., & Hughes, R. J. (2004). Improving the graduate school experience for women in mathematics: The EDGE Program. *Journal of Women and Minorities in Science and Engineering, 10*, 243–253.

Bridglall, B. L., & Gordon, E. W. (2004). The nurturance of African American scientific talent. *Journal of African American History, 89*(4), 331–347.

Burton, L. (1999). The practices of mathematicians: What do they tell us about coming to know mathematics? *Educational Studies in Mathematics, 37*, 121–143.

Case, B. A., & Leggett, A. M. (Eds.). (2005). *Complexities: Women in mathematics.* Princeton, NJ: Princeton University Press.

Cirillo, M., & Herbel-Eisenmann, B. (2011). "Mathematicians will say it this way": An investigation of teachers' framings of mathematicians. *School Science and Mathematics, 111*(2), 68–78.

Cobb, P., & Hodge, L. L. (2002). A relational perspective on issues of cultural diversity and equity as they play out in the mathematics classroom. *Mathematical Thinking and Learning, 4*(2, 3), 249–284.

Cooper, D. A. (2000). Changing the faces of mathematics Ph.D.s: What we are learning at the University of Maryland. In M. Strutchens, M. Johnson, & W. Tate (Eds.), *Changing the faces of mathematics: Perspectives on African Americans* (pp. 179–192). Reston, VA: National Council of Teachers of Mathematics.

Cooper, D. A. (2004). Recommendations for increasing the participation and success of Blacks in graduate mathematics study. *Notices of the AMS, 51*(5), 538–543.

Cromwell, A. M. (2006). *Unveiled voices, unvarnished memories: The Cromwell family in slavery and segregation, 1692–1972.* Columbia: University of Missouri Press.

Cross, W. E. (1991). *Shades of Black: Diversity in African-American identity.* Philadelphia: Temple University Press.

Datta, D. (1993). *Math education at its best: The Potsdam model.* Framingham, MA: Center for Teaching/Learning of Mathematics.

Dean, N. (Ed.). (1997). *African Americans in mathematics.* Providence, RI: American Mathematical Society and the Center for Discrete Mathematics and Theoretical Computer Science.

Dean, N., McZeal, C. M., & Williams, P. J. (Eds.). (1999). *African Americans in mathematics II.* Providence, RI: American Mathematical Society and the Center for Discrete Mathematics and Theoretical Computer Science.

DeGroot, M. H. (1989). A conversation with David Blackwell. In P. Duren (Ed.), *A century of mathematics in America (Part III)* (pp. 589–615). Providence, RI: American Mathematical Society. Originally published in 1986 in *Statistical Science 1*(1), 40–53.

Donaldson, J. A. (1989). Black Americans in mathematics. In P. Duren (Ed.), *A century of mathematics in America (Part III)* (pp. 449–469). Providence, RI: American Mathematical Society.

Donaldson, J. A. (Ed.). (1989). *Proceedings of the Eleventh Annual Meeting of the National Association of Mathematicians.*

Donaldson, J. A., & Fleming, R. J. (2000). Elbert F. Cox: An early pioneer. *The American Mathematical Monthly, 107*(2), 105–128.

Drewry, H. N., & Doermann, H. (2001). *Stand and prosper: Private Black colleges and their students.* Princeton, NJ: Princeton University Press, 2001.

Duffie, W. (2003, Summer). Alumna dismantled color barriers: A life in service to equality, education, and church. *CUA* [Catholic University Alumni] *Magazine,* 20–21.

Ellington, R. M., & Frederick, R. (2010). Black high achieving undergraduate mathematics majors discuss success and persistence in mathematics. *The Negro Educational Review, 61,* 61–84.

Falconer, E. Z. (1996). The challenge of diversity. In N. Dean (Ed.), *DIMACS Series in Discrete Mathematics and Theoretical Computer Science: Vol. 34. African Americans in mathematics* (pp. 169–182). Providence, RI: American Mathematical Society.

Flores, A. (2007). Examining disparities in mathematics education: Achievement gap or opportunity gap? *The High School Journal, 91*(1), 29–42.

Green, C. M. (1967). *The secret city: A history of race relations in the nation's capital.* Princeton, NJ: Princeton University Press.

Gutierrez, K. D., & Rogoff, B. (2003). Cultural ways of learning: Individual traits or repertoires of practice. *Educational Researcher, 32*(5), 19–25.

Gutierrez, R. (2008). A "gap-gazing" fetish in mathematics education? Problematizing research on the achievement gap. *Journal for Research in Mathematics Education, 39*(4), 357–364.

Gutstein, E. (2006). *Reading and writing the world with mathematics: Toward a pedagogy for social justice.* New York: Routledge.

Hand, V. (2010). The co-construction of opposition in a low-track mathematics classroom. *American Education Research Journal, 47*(1), 97–132.

Helms, J. E. (1990). *Black and White racial identity.* Westport, CT: Praeger.

Hermanowicz, J. C. (1998). *The stars are not enough: Scientists—their passions and professions.* Chicago: University of Chicago Press.

Herzig, A. H. (2004). Becoming mathematicians: Women and students of color choosing and leaving doctoral mathematics. *Review of Educational Research, 74*(2), 171–214.

Hilliard, A. (1995). *The maroon within us.* Baltimore, MD: Black Classics Press.

Hilliard, A. G., III. (2003). No mystery: Closing the achievement gap between Africans and excellence. In T. Perry, C. Steele, & A. G. Hilliard III, *Young, gifted, and black: Promoting high achievement among African-American students* (pp. 131–166). Boston: Beacon Press.

Houston, J. L. (1999). *The history of the National Association of Mathematicians (NAM): The first thirty years—1969–1999*. Elizabeth City, NC: NAM.

Kenschaft, P. C. (1987). Black men and women in mathematical research. *Journal of Black Studies, 18*, 170–90.

Kenschaft, P. (2005). *Change is possible: Stories of women and minorities in mathematics*. Providence, RI: American Mathematical Society.

Kessler, J. H., Kidd, J. S., Kidd, R. A., & Morin, K. A. (1996). *Distinguished African American scientists of the 20th century*. Phoenix, AZ: Oryx Press.

Ladson-Billings, G. (1997). It doesn't add up: African American students' mathematics achievement. *Journal for Research in Mathematics Education, 28*(6), 697–708.

Lefebvre, H. (1974). *The production of space* (D. Nicholson-Smith, trans). Oxford, England: Blackwell.

Leonard, J. (2008). *Culturally specific pedagogy in the mathematics classroom: Strategies for teachers and students*. New York: Routledge.

Levering Lewis, D. (1993). *W. E. B. DuBois, 1868–1919: Biography of a race*. New York: Henry Holt.

Lofton Haynes, E. (1981). Haynes-Lofton family papers [Personal and professional papers]. Catholic University of America Libraries, Washington, DC.

Lorch, L. (1996). Yesterday, today, and tomorrow. In N. Dean (Ed.), *DIMACS Series in Discrete Mathematics and Theoretical Computer Science: Vol. 34. African Americans in Mathematics* (pp. 157–168). Providence, RI: American Mathematical Society.

Martin, D. B. (2000). *Mathematics success and failure among African American youth: The roles of sociohistorical context, community forces, school influence, and individual agency*. Mahwah, NJ: Lawrence Erlbaum Associates.

Martin, D. B. (2006). Mathematics learning and participation as racialized forms of experience: African American parents speak on the struggle for mathematics literacy. *Mathematical Thinking and Learning, 8*(3), 197–229.

Martin, D. B. (Ed.). (2009a). *Mathematics teaching, learning, and liberation in the lives of Black children*. New York: Routledge.

Martin, D. B. (2009b). Researching race in mathematics education. *Teachers College Record, 111*(2), 295–338.

Megginson, R. E. (2003). Yueh-Gin Gung and Dr. Charles Y. Hu Award to Clarence F. Stephens for distinguished service to mathematics. *American Mathematical Monthly, 110*(3), 177–180.

Mehan, H., Hubbard, L., & Villanueva, I. (1994). Forming academic identities: Accommodation without assimilation among involuntary minorities. *Anthropology and Education Quarterly, 25*(2), 91–117.

Moore, J. L. (2006). A qualitative investigation of African American males' career trajectory in engineering: Implications for teachers, school counselors, and parents. *Teachers College Record, 108*(2), 246–266.

Moore Mensah, F. (2009). A portrait of Black teachers in science classrooms. *Negro Educational Review, 60*, 39–52.

Moreau, M. P., Mendick, H., & Epstein, D. (2009). What do GCSE students think of mathematicians? *Mathematics in School, 38*(5), 2–4.

Morris, J. E., & Monroe, C. R. (2009). Why study the U.S. South? The nexus of race and place in investigating Black student achievement. *Educational Researcher, 38*, 21–35.

Moses, R. P., & Cobb, C. E. (2001). *Radical equations: Math literacy and civil rights.* Boston: Beacon Press.

Murray, M. (2000). *Women becoming mathematicians.* Cambridge, MA: MIT Press.

Nasir, N. S. (2000). Points ain't everything: Emergent goals and percent understandings in the play of basketball among African American students. *Anthropology and Education Quarterly, 31*(3), 283–305.

Nasir, N. S., & Saxe, G. B. (2003). Ethnic and academic identities: A cultural practice perspective on emerging tensions and their management in the lives of minority students. *Educational Researcher, 32*(5), 14–18.

Newell, V., Gipson, J. H., Rich, L. W., & Stubblefield, B. (1980). *Black mathematicians and their works.* Ardmore, PA: Dorrance & Company.

O'Brien, V., Martinez-Pons, M., & Kopala, M. (1999). Mathematics self-efficacy, ethnic identity, gender, and career interests related to mathematics and science. *The Journal of Educational Research, 92*(4), 231–235.

O'Connor, C., Lewis, A., & Mueller, J. (2007). Researching "Black" educational experiences and outcomes: Theoretical and methodological considerations. *Educational Researcher, 36*(9), 541–552.

Pearson, W. (1988). The role of colleges and universities in increasing Black representation in the scientific professions. In M. T. Nettles (Ed.), *Toward Black undergraduate student equality in American higher education* (pp. 105–124). Westport, CT: Greenwood Press.

Pearson, W., & Pearson, L. C. (1985). Baccalaureate origins of Black American scientists: Cohort analysis. *Journal of Negro Education, 54*(1), 24–34.

Perry, T. (2003). Up from the parched earth: Toward a theory of African-American achievement. In T. Perry, C. Steele, & A. G. Hilliard III, *Young, gifted, and black: Promoting high achievement among African-American students* (pp. 1–108). Boston: Beacon Press.

Riordan, T. (2006, October 23). Massey's mentorship creates network of mathematicians. *Princeton Weekly Bulletin, 96*(7). Retrieved from https://www.princeton.edu/pr/pwb/06/1023/1b.shtml

Rucker, W. C., & Jubilee, S. K. (2007). From Black nadir to *Brown v. Board*: Education and empowerment in Black Georgian communities—1865 to 1954. *Negro Educational Review*, *58*(3 and 4), 151–168.

Scripture, E. W. (1891). Arithmetical prodigies. *Journal of Psychology, 4*(1), 1–59.

Scriven, O. A. (2006). *The politics of particularism: HBCUs, Spelman College, and the struggle to educate Black women in science, 1950–1997* (Unpublished dissertation). Georgia Institute of Technology, Atlanta.

Siddle Walker, E. V. (1996). *Their highest potential: An African American school community in the segregated South*. Chapel Hill: University of North Carolina Press.

Soja, E. W. (1989). *Postmodern geographies: The reassertion of space in critical social theory*. New York: Verso.

Southern Education Foundation. (2005). *Igniting potential: Historically Black colleges and universities and science, technology, engineering, and mathematics*. Atlanta, GA: Author.

Sowell, T. (1974). Black excellence: The case of Dunbar High School. *Public Interest, 35*, 3–21.

Stage, F. K., & Maple, S. A. (1996). Incompatible goals: Narratives of graduate women in the mathematics pipeline. *American Educational Research Journal, 33*(1), 23–51.

Steele, C. M. (1997). A threat in the air: How stereotypes shape intellectual identity and performance. *American Psychologist, 52*(6), 613–629.

Stinson, D. W. (2006). African American male adolescents, schooling, (and mathematics): Deficiency, rejection, and achievement. *Review of Educational Research, 76*(4), 477–506.

Tate, W. F. (1995). Returning to the root: A culturally relevant approach to mathematics pedagogy. *Theory into Practice, 3*, 166–173.

Tinto, V. (1993). *Leaving college: Rethinking the causes and cures of student attrition* (2nd ed.). Chicago: University of Chicago Press.

Tucker, A. (1996). Models that work: Case studies in effective undergraduate mathematics programs. *Notices of the AMS, 43*(11), 1356–1358.

Walker, E. N. (1994). *African American students' perceptions of their mathematics performance* (Unpublished master's thesis). Wake Forest University, Winston-Salem, NC.

Walker, E. N. (2006). Urban students' academic communities and their effects on mathematics success. *American Educational Research Journal, 43*(1), 43–73.

Walker, E. N. (2007). Why aren't more minorities taking advanced math? *Educational Leadership, 65*(3), 48–53.

Walker, E. N. (2009). "A border state": A historical exploration of the formative, educational, and professional experiences of Black mathematicians in the United States. *International Journal of History in Mathematics Education, 4*(2), 53–78.

Walker, E. N. (2011). Supporting giftedness: Historical and contemporary contexts for mentoring within Black mathematicians' academic communities. *Canadian Journal for Science, Mathematics, and Technology Education, 11*(1), 19–28.

Walker, E. N. (2012a). *Building mathematics learning communities in urban high schools.* New York: Teachers College Press.

Walker, E. N. (2012b). Cultivating mathematics identities in and out of school and in between. *Journal of Urban Mathematics Education, 5*(1), 66–83.

Wiggan, G. (2007). Race, school achievement, and educational inequality: Toward a student-based inquiry perspective. *Review of Educational Research, 77*(3), 310–333.

Wilmot, N. (2003). *An oral history conducted with David Blackwell, 2002–2003.* Berkeley, CA: Bancroft Library, University of California.

Winston, M. R. (1971). Through the back door: Academic racism and the Negro scholar in historical perspective. *Daedalus, 100*, 678–719.

Young, R. V. (1998). *Notable mathematicians from ancient times to the present.* Detroit, MI: Gale Publishers.

INDEX